城市设计研究
步行与干道的合集

URBAN DESIGN RESEARCH
PEDESTRIAN PARADISE
ON ARTERIAL ROAD

【著】孙彤宇 许凯
Sun Tongyu　Xu Kai

同济大学出版社

沿路发展的城市
上海浦东,1990

URBAN DEVELOPMENT ALONGSIDE ARTERIAL ROAD
PUDONG, SHANGHAI, 1990

关于本书

机动车时代，城市干道在城市空间结构及城市发展、演变中起到了举足轻重的作用，城市几乎无法离开干道而存在。与此同时，城市的步行系统已经退化到只剩下依附于干道的人行道系统，已基本无法成为日常城市生活的公共场所。要解决机动车时代的步行环境问题，首当其冲的便是提升依附于城市干道的人行道的空间品质以及打造与此相关的公共空间体系。本书经过理论研究、实践案例调研、具体案例的设计，展现给读者的是机动车时代实现"步行者天堂"的潜在可能性——步行与干道的合集。

About This Book

In the Motor Age, arterial road plays an important role in contemporary city's spatial structure and its development. Meanwhile, the pedestrian system in contemporary cities has degenerated to affiliated infrastructure to arterial roads and thus has lost its role as public space serving people's daily life. To improve a city's walkability, the primary challenge is to improve spatial quality of sidewalks alongside the arterial roads and further to establish public space system that is based on it. This book introduces research on theory, case study and urban design proposal for selected urban areas, and shows to the readers how a "pedestrian paradise" can be true in contemporary cities.

不适合步行的干道空间
北京CBD,2005

UNPLEASENT PEDESTRIAN SPACE
ALONGSIDE ARTERIAL ROAD
BEIJING CBD, 2005

关于研究课题和研究生国际设计课程

针对机动车时代步行环境的不足，在国家自然科学基金项目"TOD模式下步行系统与城市公共空间及交通耦合模型研究"的资助下，课题组展开了多种形式的学术研究和设计实践。在此基础上，将研究初步成果中得到的有关结论和解决原则等作为指导思想，编制同济大学硕士研究生国际设计课程"步行者天堂"，以进行更加深入的探索和设计实践。连续两年的课程共有28位中国学生和15位来自德国、法国、西班牙、奥地利、芬兰等国的国际双学位学生和交换生参与，邀请了来自奥地利、比利时、美国、韩国、荷兰等国著名建筑院校的教师参与指导和评图，还邀请了上海市规划局与虹口区规划局的专家共同研讨。课程取得了丰硕的成果，并进行了两次成果展览，大部分作品作为本书的案例展示。本书的出版是科研与教学良好结合和互动的有益探索。这种探索还将继续，我们期待在不远的将来向读者展现更为丰厚的研究成果。

About the research project and international design course

Aiming at the improvement of pedestrian quality in the Motor Age, and supported by NSFC project "Study on the Coupling Model between Pedestrian System and Urban Public Space and Traffic under TOD Mode", the research team conducts various forms of research, teaching and design practice. The findings and principals of the research led to the design courses "Pedestrian Paradise" for international master students in Tongji University, in which students were motivated to come up with urban design ideas for specific cases in city. In 2 courses that took place in 2013 and 2014, 28 Chinese students, 15 foreign students (from Germany, France, Spain, Austria and Finland) and many international professors (from Austria, Belgium, Korea and Netherlands) had taken part in. The works of students have been exhibited in several occasions. Many of them are shown in this book.

回归步行的干道空间
维也纳玛利亚大街，2015

ARTERIAL ROAD
RETURNING TO PEDESTRAIN FRIENDLY PLACE
MARIAHILFERSTRASSE, VIENNA, 2015

回归步行的干道空间
韩国首尔清溪川,2008

ARTERIAL ROAD
RETURNING TO PEDESTRIAN FRIENDLY PLACE
CHEONGGYECHEON, SEUOL, 2008

作者简介 Biography

孙彤宇 | 同济大学建筑与城市规划学院博士，德国柏林工大和斯图加特大学访问学者。现任同济大学建筑与城市规划学院副院长、教授、博士生导师，高密度区域智能城镇化协同创新中心特聘教授，同济—柏林工大智慧城市国际联合实验室中方负责人，同济—维也纳工大双学位项目负责人。主要研究领域为城市设计及建筑设计理论与方法，主持多项国家和省部级科研项目。同时也是活跃的从业建筑师，是同济大学建筑设计研究院（集团）有限公司都市建筑设计院四所所长和主创建筑师，有多项设计获国家和地方建筑设计奖。

Prof. Dr. Sun Tongyu obtained his doctor degree in Tongji University in Shanghai. He was the visit scholar in Technical University of Berlin in 2003, and in in Stuttgart University in 2008. He is now the vice dean of the College of Architecture and Urban Planning, Tongji University, honorary professor of Intelligent Urbanization Co-creation Center for High Density Region, the director of Tongji-TU Berlin International Joint Laboratory of Smart-City. His main research field is urban design and architecture design. Being the director of design division four in Tongji Design Institute,He also actively practices in architecture design and urban design, and gained some design awards.

许凯 | 同济大学建筑与城市规划学院硕士，维也纳工大建筑学院博士，师从Klaus Semsroth教授。现任同济大学建筑与城市规划学城市设计团队副教授，同济—维也纳工大双学位项目协调人。主要研究领域是城市设计与产业空间的规划。作为上海尤根建筑设计的创办者，从事建筑设计和城市设计方面的工作。

Dr. Xu Kai is an associate professor in Tongji University of Shanghai. He obtained his doctor degree in University of Technology,Vienna. Prof. Klaus Semsroth was his tutor professor. Dr. Xu's research areas include urban design and industrial space planning. Being the founder of Jugend Architecture, he also actively practice in architecture design and urban design.

序

探求城市的适宜复杂性

案头上放着《步行与干道的合集》的清样，书名就令人感兴趣，关于城市设计普遍为人们接受的概念是人车分流、步行街、步行街区、广场、场所等，而孙彤宇教授和许凯副教授的这项研究提出了一个全新的概念，从另一个角度为我们展现了城市空间的复杂性和城市设计的新视角。

城市设计的概念最早于1943年出现在英国规划师和建筑师艾伯克隆比和约翰·亨利·福肖为战后伦敦规划的思想中，并在第二次世界大战后成为一种专业活动，城市设计是设计并重塑城市的重要手段和过程。美国城市设计理论家凯文·林奇坚持在人类生存环境中应有视觉上的可识别的秩序，他所主张的形式的形象性和可辨识性成为规划师和建筑师们最关注的与传达意义有关的元素。意义存在于路径、边缘、节点、区域以及地标的特殊性中。这一类城市设计理论注重形式，注重秩序和图底关系，注重城市空间及其形态。在这种思想指导下，人们以怀旧的态度指责现代城市的非人性，推崇中世纪城市的有机生成和非规整空间，复杂、综合但协调一致的城市空间。

实际上，城市是一个极为复杂的范畴，所涉及的方面相当广泛，而且所有的这些问题都仍然处于持续不断的进化和演变的过程之中。城市几乎涉及所有的领域和所有的人，让人们有充分想象、创造和表现的空间。城市也是混沌的集合体，表现为城市功能的混合，城市中的秩序与混沌的结合，城市空间的多元。日本建筑师筱原一男在《混沌与机器》（1988）中提出了以混沌为中心的美学理念，他用混沌理论来看待城市和建筑，他认为："混沌是城市的基本特征。" 城市在本质上是多元的分形结构。传统城市具有明显的分形特征，而自20世纪以来的现代建筑运动形成的城市推崇反分形的几何性类型学。在生态学的理论中，过于复杂的社会系统会难以继续有效地运行，当社会复杂性超过最大限度，或者最适宜的程度时，就会开始经历严重的环境问题，丧失应变能力，并开始衰退。因此，适宜的复杂性是所有复杂系统应当寻求的发展目标。

孙彤宇教授和许凯副教授的研究探索的正是城市的适宜复杂性，涉及最为普遍而又最受诟病的城市主干道现象。正如作者在书中开宗明义就指出的："机动车时代，城市干道在城市空间结构及城市发展、演变中起到了举足轻重的作用，使得城市几乎无法离开干道而存在。"由于干道不为人们称道的交通性特征，目前关于城市设计的论著几乎很少涉及城市干道。实际上，我们耳熟能详的那些作为经典的城市主干道和商业大街都不是纯粹的单一功能的步行街，例如巴黎的香榭丽舍大道和里沃利大街、罗马的主街、柏林的菩提树下大街和选帝侯大街、纽约的第五大道和麦迪逊大道、东京的银座和表参道、维也纳的环城大道、伦敦的牛津大街、北京的长安街、上海的淮海路和南京西路等等概莫能外。试想一下，这些大街如果只是单纯的交通干道，或者反过来只是单纯的步行街的结果会是什么，也许这些大街正是由于车水马龙和人流的熙熙攘攘才成其为世界级的大街的，而且几乎所有城市的主街都是这个情况。为此，孙彤宇教授和许凯副教授创造了一个术语"合集"，形象而又生动地陈述了这一交通干道与步行共生的现代城市特征。大多数城市在历史上都是沿干道发展起来的，无论这种干道是走毛驴的、走马车的或是走汽车的，甚至走火车的，沿着干道形成市集，形成城市。

《步行与干道的合集》以建筑师的艺术化的图式语言和精炼的说明向我们展示了城市干道的空间关系，不仅有理论分析，而且有许多上海的实践案例，这些案例几乎就发生在建筑师的眼皮底下，表现出建筑师和学者超越校园空间的制约，参与改造城市空间的成果。

《步行与干道的合集》告诉我们，建立在复杂城市空间上的理想是可以成为现实的，也告诉我们要关注和研究那些往往会视而不见的城市现象的必要性，把握适宜的复杂性，对于城市更新和城市设计，建筑师是可以大有作为的。

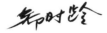

中国科学院 院士、法国建筑科学院 院士
同济大学建筑与城市规划学院 教授、博导
2016年11月30日

INTRODUCTION

To Explore The Appropriate Complexity of City

The name of this book *Pedestrian Paradise on Arterial Road* interests me by a brand-new concept which is very different from prevailing concepts normally focusing on pedestrian system separated from vehicle system, pedestrian street, pedestrian urban blocks, plazas and places. The concept to integrate pedestrian system with arterial roads reveals a new perspective of urban design, as well as the fact how complex the urban space can be.

Urban design gained its ground in "The Greater London Plan" (1944) by Sir Leslie Patrick Abercrombie and John Henry Forshaw and since then became a profession being treated as important method and process to design and reshape city. American theorist Kevin Lynch believes that the human environment shall possess recognizable visual order, therefore urban planners and architects must take recognizable urban forms with strong image as most important elements in city to convey meaning. Influenced by this theory, the inhumane images of modern city are criticized, while those organic, irregular, complex but harmonious spaces in medieval cities are praised.

However, city is such a complex system which covers numerous aspects which are still in process of evolution and change. City is a space for all people to imagine, create and perform in. City is also a chaotic integration of elements, demonstrated by mixture of uses, mixture of order and disorder, and mixture of polybasic spaces. In his book *Chaos and Machine*, Japanese architect Kazuo Shinohara brings forward the aesthetics based on chaos, which treats city and architecture by Chaos Theory and leads to the conclusion that "chaos is the core identity of city". Various researches have revealed the fractal characteristics of city. These strong fractal characteristics of traditional cities are erased and replaced by simple geometric order during Modernism Movement. On the other hand, theory of Ecology reveals that a sustainable social system cannot be over complex, which might lead to severe social problem and loss of evolution capability. Therefore, an appropriate level of complexity shall be the target of all complex systems' evolution.

The research by Sun Tongyu and Xu Kai takes exactly the appropriate complexity of city as its research domain, focusing on arterial road in city, which is the element so important to city and so widely criticized. As is already stated in the very beginning of this book, "in the Motor Age, arterial road plays an important role in contemporary city's spatial structure and its development". However, due to the low pedestrian quality as a result of sacrifice to its traffic function, it is not under focus of main-stream academic research. In fact, many well-known arterial roads or commercial roads in the world, many of them being treated as "classics", are not purely pedestrian road. The famous Avenue des Champs - Elysées and Rue de Rivoli in Paris, Unter den Linden and Kurfürstendamm in Berlin, the 5th Avenue and Madison Avenue in New York, Ginza and Omotesando in Tokyo, Ringstrasse in Vienna, Oxford Street in London, Chang'an Road in Beijing, as well as Huaihai Road and West Nanjing Road in Shanghai, are extinguished cases among them. The prosperity of these roads lies exactly in that they are not "pedestrianized", instead, a integrative ecology of pedestrian environment and traffic function is founded. In the research by Sun Tongyu and Xu Kai, the term "integration" is coined to demonstrate such situation. Indeed, most cities had been developed based on arterial roads, either they are for animals, carriages, automobiles or even for trains, which grow to be places for market and later the whole city is born based on them.

Pedestrian Paradise on Arterial Road elaborates to readers the spatial characteristics of arterial roads by diagrammatic illustrations and refined texts. Many of the cases shown in this book are located in Shanghai, which shows the author's effort as urban designer to participate in the urban development process of Shanghai.

This book shows to us, a dream built on complex space system can be realized, through revealing the undiscovered urban phenomenon and its complexity. Urban design and urban regeneration can be well achieved by architects with efforts to pursue appropriate complexity in city.

Prof. Dr. Zheng Shiling
Academician of Chinese Academy of Sciences
Academician of French Academy of Architecture
Nov. 30, 2016

目录
CATALOGUE

关于本书
ABOUT THIS BOOK

关于研究课题和研究生国际设计课程
ABOUT THE RESEARCH PROJECT AND INTERNATIONAL DESIGN COURSE

作者简介
BIOGRAPHY

序
INTRODUCTION

前言 机动车时代的挑战与机遇 ······ 18
CHALLENGES & OPPORTUNITIES IN MOTOR AGE

1 城市扩张与干道的出现 ······ 20
URBAN EXPANSION AND ARTERIAL ROADS

2 干道主导下的城市空间结构 ······ 26
URBAN FORM DOMINATED BY ARTERIAL ROADS

3 "中心"还是"边缘":干道的命运转折 ······ 38
"CENTRAL" OR "PERIPHERAL": THE TURNING POINT OF ARTERIAL ROADS

4 城市沿干道发展的空间特征 ······ 42
SPATIAL CHARACTERISTICS OF AREAS ALONGSIDE ARTERIAL ROADS

5 城市空间意象的变异 ······ 52
DUAL URBAN IMAGES IN MOTOR AGE

6 城市空间可步行性研究的基础和进展 ······ 62
RESEARCH BACKGROUNDS AND PROGRESS ON URBAN WALKABILITY

7

干道与步行的合集: 愿景与行动 ···················· 68
PEDESTRIAN PARADISE ON ARTERIAL ROAD: VISION AND ACTION

8

城市设计的实践:
四平路沿线步行系统优化 ···················· 80
URBAN DESIGN PRACTICE:
TO IMPROVE PEDESTRIAN ENVIRONMENT ALONGSIDE SIPING ROAD OF SHANGHAI

相关城市设计提案 URBAN DESIGN PROJECT

(1) 五角场 ···················· 90
    WUJIAOCHANG
(2) 四平路—国权路 ···················· 110
    SIPING ROAD-GUOQUAN ROAD
(3) 四平路—中山路 ···················· 118
    SIPING ROAD-ZHONGSHAN ROAD
(4) 同济校门 ···················· 126
    MAIN GATE OF TONGJI CAMPUS
(5) 四平路—赤峰路 ···················· 144
    SIPING ROAD-CHIFENG ROAD
(6) 四平路—大连路 ···················· 172
    SIPING ROAD-DALIAN ROAD
(7) 四平路—曲阳路 ···················· 194
    SIPING ROAD-QUYANG ROAD
(8) 四平路—临平北路 ···················· 200
    SIPING ROAD-LINPING NORTH ROAD
(9) 四平路—海伦路 ···················· 206
    SIPING ROAD-HAILUN ROAD
(10) 吴淞路—海宁路 ···················· 222
    WUSONG ROAD-HAINING ROAD
(11) 吴淞路—外滩 ···················· 234
    WUSONG ROAD-THE BUND

参考文献 ····················250
REFERENCES

后记 ····················254
EPILOGUE

机动车时代的挑战与机遇

CHALLENGES & OPPORTUNITIES IN MOTOR AGE

　　20世纪见证了城市交通方式的巨大改变，机动车代替步行成为城市出行的主要方式，城市也因此获得了支配更庞大空间规模的能力。当代城市的发展和道路基础设施紧密地联系起来，物流、人流、能量流、信息流搭载在机动车干道上被输送至各个城市功能终端。尺度巨大、功能单一的城区出现，它们取代传统功能混合的城区，依靠机动车道路来实现功能上的连接。在享受新的技术手段带来的便利同时，人们享受着更宽广的空间，对空间认知的距离也不断扩展；但是，这个典型的20世纪城市梦也有不那么完美的地方：机动车干道上的交通堵塞成为大城市人们的梦魇，城市干道两侧的恶劣环境无法适宜人们的步行活动。在功能单一化的城区中，城市生活不再繁盛。传统步行城市中那些充满活力并充当着社会融合场所的城市空间网络在我们的时代正在逐渐消失，这是机动车时代城市发展面临的一个重大问题。

　　不可否认的是，城市干道已经成为当代城市不可缺少的组成部分。在城市扩张阶段，它是城市扩张的有力工具并奠定了城市的基本骨架；今天，它们却成为城市中令人不快的场所。对如何看待这个问题，有没有解决的方法，是本书将讨论的主要内容。

Motor Vehicle became main means of urban transportation in 20th century and gave city great capacity to dominate space in much larger geographical area. Today urban development has been closely associated with development of arterial traffic roads as urban infrastructure to carry material, people, energy and information to each functional terminal in the city. It is also during this period when huge urban zones with mono function appear (such as residential zones, commercial zones, industrial zones and so on), replacing the "old-fashioned" mix-use areas. They are born to be linked by arterial traffic roads. People enjoy the convenience brought by new transportation means while their recognition of space and distance is much expanded. However, such typical 20th Century Urban Dream has its imperfect side. Traffic jam is among the nightmares of 20th Century, while the urban environment alongside traffic road becomes so unfriendly for people. In those mono-function zones, urban life no more prospers. Urban space, which is filled with urban vitality and acts as place for social integration in traditional cities, disappears gradually in our era. This might be treated as one of the main challenges for city in the Motor Age.

What cannot be denied is that arterial roads are indispensable and inseparable part of most contemporary cities. During rapid expansion period of city, the planning of such roads has become a major instrument of urban planning and furthermore defined the basic spatial structure of the city. Today most of these roads and their surrounding areas are regarded as unpleasant places in city. How we shall look at this problem theoretically and practically, as well as how to find out a solution in terms of urban design, is the major content of this book.

1 城市扩张与干道的出现
URBAN EXPANSION AND ARTERIAL ROADS

图1a 豪斯曼巴黎扩张计划中的道路规划(1853)
Fig. 1a Arterial roads in Haussmann Paris (1853)

图1b 19世纪末维也纳的城市扩张规划
Fig. 1b Urban expansion of Vienna in the end of 19th century

图1c 19世纪中叶的巴塞罗那
Fig. 1c Urban expansion of Barcelona in the middle of 19th century

空间扩张是工业革命以后城市发展的普遍趋势，是激增的城市人口和同样激增的人均用地共同作用的结果，而交通技术的发展和方式的改变，是这个趋势得以形成的强大支撑。19世纪后半叶，欧洲大城市里公共交通的出现让城市空间进入第一次大规模扩张。巴黎（1853）、维也纳（1857）和巴塞罗那（1859）野心勃勃的扩张都是以有轨电车和后续地铁与城铁系统的架设为基础的（图1a—图1c）。20世纪初私人汽车交通的出现推动了城市空间扩张的第二轮热潮。如果没有私人汽车，美国那些蔓延的低密度城市也许根本不会产生（图2）。以前人们必须生活在两公里直径的城区范围内，因为他们支配空间的方式是步行，而现代人支配空间的方式让他们可以长途迁徙去一个地方工作，却在另一个地方居住。

城市干道作为当代城市的要素，正是在这样的语境下产生的。与上述城市发展阶段相对应，城市干道的发展也经历了两个阶段：最初干道服务于有轨电车、马车和步行；私人小汽车出现之后，城市干道演变为主要服务于私人汽车了。这两个时代城市干道的巨大区别是，前者服务的交通方式速度都比较低，能够与步行活动较好地融合，因而城市干道往往成为城市中重要的活动场所，如巴黎的那些林荫大道和维也纳的指环路（图3—图6）。之后时代的交通干道上城市活动的内容因为机动车造成的安全、噪声和尾气影响被极大伤害。

从城市空间扩张的角度看，无论在上述的哪个阶段，城市干道都无疑是城市空间扩张的有力工具。康泽恩地理学的研究告诉我们，在组成城市空间系统的三个要素中（建筑、地块和道路），道路是最不容易被改变的，因为它涉及众多产权和利益体的平衡关系（Conzen，1990）。斯蒂芬·马歇尔的研究则揭示了道路网生长的过程中干道的基础作用（Stephen Marshall，2014）。反过来说，扩张过程中城市干道的确定在一定程度上为新的空间结构打下了基本框架。19世纪后半叶，巴黎、维也纳和巴塞罗那的干道,规划和建设几乎完全改变了当时城市的空间结构，为城市扩张打通了空间通道并确立发展的新逻辑(图7)。20世纪初期关于城市扩张的各种乌托邦模型中，干道也作为主要的空间要素被首先计划，其优先程度甚至

图2 美国都市变迁中的主干道
Fig. 2 Arterial road in urban sprawl

图3 机动车时代以前的巴黎香榭丽舍大街
Fig. 3 Avenue des Champs-Elysees before the Motor Age

图4 巴黎香榭丽舍大街上的公共生活
Fig. 4 Public life on Avenue des Champs-Elysees

图5 维也纳指环路上的公共生活
Fig. 5 Public life on Ringstrasse of Vienna

图6 维也纳的指环路与公园
Fig. 6 Ringstrasse and parks in Vienna

超过土地利用和开放空间的规划。很难想象，霍华德的田园城市如果少了那些围绕主城区和卫星城的环路和辐射路（图8），或者马塔的线形城市少了作为主动脉的主干路会是怎么样的。

现实中的城市扩张往往以干道规划和首先被建设为起始（无论是干道的一段还是完整的道路，在投资充裕的情况下，甚至整个干道网都会先期建设），随后，城市的建设则会沿着干道发生。当干道两侧地块的建设趋于饱和之后，会接着向纵深方向扩张，从而形成完整城区。对于一些历史比较短的城市而言，这种线性发展的特点尤为明显，整个城市的形态基本沿着城市干道展开，例如巴西利亚和深圳。对于那些发展历史长一些的城市而言，它们往往有一个很强的核心，在机动车时代以前就形成，但在当代进行扩张的语境下，城市先是沿着以老城为圆心的几条干道辐射状向外扩张，逐渐将城市空间格局撑开；接下去的发展往往会出现以老城边缘为第一层环路，并逐次向外环绕第二层、第三层环型干道。后者显示出一种比辐射型干道更雄心勃勃的扩张姿态，因为这种发展不仅是"撑开"，还显示了强烈"占满"的意图，前者例如维也纳以老城为核心向外辐射的几条大道，后者例如北京陆续规划的六条环道。当然，现实中的城市扩张往往根据各自城市的实际状况进行，综合上述两种模式，例如上海已经形成的复杂交错的辐射型、网格型和环路型干道路网情况。

Spatial expansion is the general tendency of urban development of almost all cities after Industrial Revolution. The expanded population and expanded land use per capita serve jointly as driving force, while the renovation of transportation technology being the strong support for this tendency. The first urban expansion period took place in the second half of 19th century in Europe right after the motorized mass public transportation (tramways and later subways) had emerged, taking the example of Paris (1853), Vienna (1957) and Barcelona(1859)(Fig.1). The second period of expansion took place when the motorized personal transportation method emerged. Urban sprawl in many American cities would be hard to imagine without the personal automobile(Fig.2). In contrast to premodern people who had to live within an urban scope of less than 2 kilometers diameter due to pedestrian-based behavior, people in city today can live and work in city which is much broader spatially, giving the condition they can move in much larger distance by motorized transportation.

In such context, arterial traffic road as an important urban element in modern city emerged. In correspondence to the aforementioned two periods of city's urban expansion, the development of arterial traffic road experiences also two periods. It served in the first period mainly for tramways, carriages and pedestrians, and then in the second period mainly for personal automobile. The big difference between arterials roads during these periods is that, in the former one the motorized transportation can be well compatible with pedestrian behavior and thus they became important urban space in city (such as the Ringstrasse in Vienna and those famous Boulevards in Paris)(Fig.3-6), while in the later one motorized transportation (personal automobile) excludes pedestrian behavior due to safety, noise and emission problems caused by it.

If we look at the things from perspective of urban expansion, no matter during which periods it is, the arterial roads are doubtlessly an important planning instrument for urban expansion. The Geographical research by Conzen told us, within the three main elements of urban system (building, land and street), street is the most difficult one to change, because its change may involve many properties and stakeholders in the city and thus break the balance(Conzen,1990). Stephen Marshall's research reveals the growing mechanism of road net, in which arterial road play as basis (S·Marshall,2014). On the other hand, when an arterial road is planned and developed in the initial period of urban expansion, it means the primary spatial structure of future urban area is laid down basically. During the urban expansion of Vienna, Paris and Barcelona in the sec-

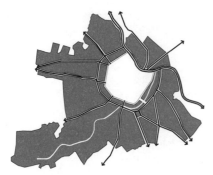
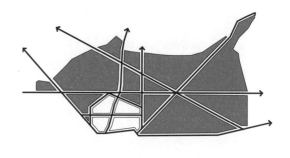

（a）19世纪维也纳城市扩张规划中的城市干道
Arterial roads in Vienna in the 19th century

（b）19世纪巴塞罗那城市扩张规划中的城市干道
Arterial roads in Barcelona in the 19th Century

图7 19世纪欧洲城市扩张中的干道
Fig7. Arterial roads in European cities in the 19th century

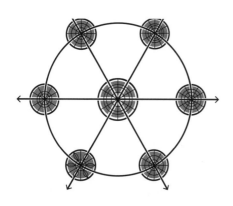
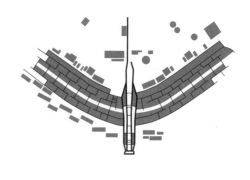

（a）田园城市规划模式中的城市干道
Arterial roads in the Garden City

（b）巴西利亚规划模式中的城市干道
Arterial roads in Brazilia

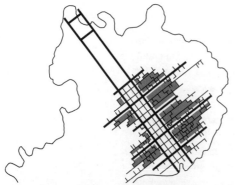
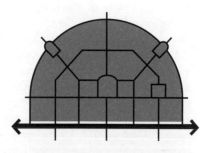

（c）东京湾规划规划模式中的城市干道（丹下健三）
Arterial roads in Tokyo Bay Planning (Kenzo Tange)

（d）TOD模式中的城市干道（P.卡尔索普）
Arterial roads in TOD model (P. Calthorpe)

图8 几种当代城市模型中的干道
Fig8. Arterial roads in some urban models of modern city

ond half of 19th century, the planning of arterial roads had played an important role (Fig. 7). In those many utopian urban models invented in the beginning of 20th century, the planning of arterial roads was also taken as main planning instrument, even prior to the planning of land use and public space. It would be very hard to imagine the Garden City Model by Ebenezer Howard or the Band City Model by Mata, without being aware there are the arterial roads as the aortas (Fig. 8).

Urban expansion in reality is usually initiated with planning and development of arterial roads, no matter it is in form of entirety or fragment, sometimes if the funding is abundant the whole arterial road net is laid down. And then the city development will follow, taking the arterial roads as aortas. The next stage of development will go perpendicularly direction of the road only after the areas alongside the road are completed. For those cities which have relatively short history of development, such linear-shaped development model is more obvious, such as Brazilia and Shenzhen. And for those cities with longer history, they normally have a stronger core area which took its form in pre-motor-vehicle age, and started to expand following arterial roads in radiation direction. The next stage of expansion will be taking the old city core's edge as first ring road, and then being followed by the second and the third. The later model of expansion show much ambitious intension as it aims not only in expansion itself but also in occupation of the space being expanded. Vienna's first ring road and the radiation roads such as Thaliastrasse, Alserstrasse and Landstrasse may belong to the former model, and Beijing's six ring roads may belong to the later. Certainly, many other cities take its existing situation as basis and use a model being hybrid of the two mentioned before. For example, Shanghai's arterial road net might be treated as a hybrid of radiation, grid and rings.

2 干道主导下的城市空间结构
URBAN FORM DOMINATED BY ARTERIAL ROADS

　　传统的步行城市里以空间集聚为基础的扩张与当代城市的沿路扩张形成两种截然不同的城市空间结构模式。我们可以拿上海中心城区的浦西部分和浦东部分作为两个典型来进行比较。前者的主要路网结构在机动车时代之前已经形成，尽管在进入机动车时代后进行过适应机动车的改造，但其适合步行的基本特点仍比较明显。后者是在20世纪90年代后期才被规划建设，城区的规划基本上是由干道主导的。这两个城区呈现出截然不同的空间结构特点，归纳为如下基本指标和特征（图9—图11）。

■ 开发强度指标

　　这个指标指的是一个城区所容纳的建筑物总量与城区面积的比率。很有意思的是，尽管以下的诸多指标都显示出两个城区空间结构上的极大不同，但在开发强度这个指标上却是相似的。这显示了在容纳相似人口和满足城市发展需求的情况下，城市空间结构的模式差异。

■ 地块建筑密度指标

　　对该指标的判定可以直接从城市黑白图上面得出直观的结论。从每个独立地块内部的组织方式上看，如果仅比较住宅部分用地的话，浦西部分的地块建筑覆盖率普遍高于40%，对于很多里弄住区而言这个指标甚至高于60%。如果算上街道的占地，这些区域的平均密度也高于30%，呈现出致密的肌理特点；浦东部分住宅区的地块覆盖率则低于30%，算上道路占据的大量用地，这些区域的平均密度非常低，因而肌理疏松。从地块之间的关系上看，前者的地块关系非常紧密，原因是用于区分地块的道路路幅较小，而地块两侧的界面也相对连续；后者则相对独立，宽路幅的机动车干道导致地块之间距离较远，地块周边两侧的建筑界面不连续也进一步导致地块关系模糊。

Arterial-road-based urban expansion led to an urban form model which is definitely different from traditional city's urban form which develops based on pedestrian behavior. We can take Shanghai's Puxi Area and Pudong Area as two typical cases for comparison. The road net for the former one took its shape already in pre-motor-age, although it had been restructured for motorized traffic during rapid development period of Shanghai, its character as pedestrian-friendly area is still outstanding. The later one was planned and developed after later 1990s, in which the arterial roads act as dominating elements. The two cases represent very different structural quality, which are summarized with following indexes and characteristics (Fig.9-11).

■ Development intensity

This index indicates the ratio of total amount of building floor areas by the land these areas occupied. Interestingly, the two cases exhibit similar level of development intensity. Such similarity showcases the urban form may vary even if they are conforming similar demand of population and urban development.

■ Density of building

This index can be visualized from figure-ground Plan of the urban area, in which buildings are painted black. In Puxi Area the building density is relatively high which makes the urban fabric very compact and organic. Taking residential areas for example, building density in most residential areas are higher than 40%, and in some areas where traditional housing form Linong was built, building density may be up to 60%. Meanwhile, the building density is much lower and it makes the urban fabric much uncompact. Building density of most of the residential area is less than 30%. From perspective of the relation between different building blocks, it is much closer and compact in Puxi Area due to narrower street and more continuous building surface in blocks' boundary. In Pudong Area, the wide traffic roads (some of the arterial roads are wider than 100 meter) make the wide distance and undefined connection between blocks .

■ 道路宽度与密度指标

　　相较浦东，浦西部分城区步行时代产生的路网结构中，道路本身的断面更小，路网的密度更大，路网形态也更复杂和有机，层级丰富。浦东部分城区的道路断面大，路网密度急剧下降，路网形态比较规则，符合机动车通行的需求。浦西中心城区道路在近期被持续地进行改造以符合机动车通行的特点，如道路拓宽和架设内环高架路。这些改造在原有的道路网结构上添加了一个供机动车使用的层级，令浦西部分的城市路网结构呈现出一种更加复合的状态，与浦东部分城区道路网的单一结构形成强烈的反差。

■ 功能混合度指标

　　步行时代的各种城市功能如居住、商业、工作、文化设施、公共绿地等，往往被组织在较小的、步行可及的城区范围内，以保证人居的便利和各种城市活动在步行条件下得以展开。然而，机动车时代将人们的可达范围扩大，令城市功能可以在更大的范围内进行组合。从上海两个城区的比较上可以看出，浦西部分城市地块尺度比较小，不同功能的地块相互组合，甚至在单一地块之内还能够形成混合的功能，浦东部分城区的城市地块尺度则大得多，每个地块功能比较单一，相同功能的地块也往往连绵成片。

■ 公共空间分布

　　浦西部分城区的公共空间举世闻名，其主要构成元素是具有良好步行体验的街道以及小型的开放空间和公园——这些元素之间往往能形成比较好的网络关系，保证公共活动在连续的体验中展开。然而，在一些后来被拓宽的机动车干道处，公共空间网络断裂。在浦东部分的城区，由于机动车道路目前并不具有传统街道的宜步行特征，两侧空间体验较差，所以这些道路并未成为公共空间网络的组成元素。现状公共空间主要由大型广场和绿地组成，布局上难以相互连接，呈现集中分布、彼此隔离的状态。

■ Width of roads and its density

The streets in Puxi Area was basically formed before the Motor Age, they are much narrower and more densely laid. The form of street net is complex and organic with very rich layers. In Pudong Area, there is hardly anything to be named "street net". The transportation system is composed of traffic roads which are wider in its section and are more loosely laid. The net's form is regulated in grid shape, which is meant to meet demand of motor traffic. An interesting fact is that the street net in Puxi Area has experienced restructuring process continuously up to today to be more suitable for motor traffic, for example many streets have been widened to be arterial traffic roads (such as Remin Road, Haining Road, Fuxing Road etc.) and later the elevated road were planned to be most dominating traffic structure inner city area. The newly added structure which serves the motor traffic adds a new layer on the original street net, which jointly leads to a very composite state of transportation system. This forms somehow a very big contrast to the Pudong Area transportation system which is much more rigid.

■ Urban function mixture

Before the Motor Age, different urban functions such as residential, commercial, facility, cultural and parks must stay in a relatively small area which is accessible by pedestrian to ensure convenience of living and the possibility of other urban activities. In the Motor Age the distance between different functions can be much bigger. In the comparison of the two cases in Shanghai, we may discover that in Puxi Area the scale of blocks are much smaller and they usually has different functions. Sometimes mixture of functions takes place in the same block. While in Pudong Area the building blocks are much larger and normally planned with single function. Blocks with same urban function always stay together.

■ Public space distribution

The ways of public space distribution differ in two cases. Puxi Area is world famous for its public space, which is composed of pedestrian-friendly urban street as well as small-scaled community public space and parks. These elements are well linked that ensures the public activities to be conducted in a continuous net. Such net is only broken in certain places where the arterial traffic roads are built. In Pudong Area, traffic roads fail to provide pedestrian friendly environment and thus they don't become elements of public space. The existing public space is composed of big-scaled plazas and greenery. Due to the absence of small-scaled streets as linkage, these elements are very fragmentary.

图9 上海浦西与浦东的城市肌理
Fig. 9 Urban Fabric of Puxi and Pudong in Shanghai

图10 上海浦西与浦东的路网结构
Fig. 10 Road Net of Puxi and Pudong in Shanghai

浦西—人民坊
开发强度（容积率）：2.9
综合建筑密度：35.9%
道路宽度：8~18米
每平方公里用地里的道路长度：13.2公里

Puxi - Renmin fang
Development Intensity (FAR) : 2.9
Average Building Coverage : 35.9%
Width of Roads : 8~18m
Total Length of Road per SQKM : 13.2km

浦西—田子坊
开发强度（容积率）：2.4
综合建筑密度：39.4%
道路宽度：5~8米
每平方公里用地里的道路长度：15.2公里

Puxi - Tianzi fang
Development Intensity (FAR) : 2.4
Average Building Coverage : 39.4%
Width of Roads : 5~8m
Total Length of Road per SQKM : 15.2km

浦西—外滩
开发强度（容积率）：2.7
综合建筑密度：40.5%
道路宽度：8~18米
每平方公里用地里的道路长度：10公里

Puxi - The Bund
Development Intensity (FAR) : 2.7
Average Building Coverage : 40.5%
Width of Roads : 8~18m
Total Length of Road per SQKM : 10km

浦西—武康路
开发强度（容积率）：2.1
综合建筑密度：28.1%
道路宽度：4~12米
每平方公里用地里的道路长度：11.2公里

Puxi - Wukang road
Development Intensity (FAR) : 2.1
Average Building Coverage : 28.1%
Width of Roads : 4~12m
Total Length of Road per SQKM : 11.2km

图11　上海浦西与浦东的城市空间结构对比
Fig. 11　Urban Form in Puxi and Pudong in Shanghai

浦东—东方明珠
开发强度（容积率）3
综合建筑密度：13.4%
道路宽度：24~100米
每平方公里用地里的道路长度：6公里

Pudong - The oriented pearl tower
Development Intensity (FAR) : 3
Average Building Coverage : 13.4%
Width of Roads : 24~100m
Total Length of Road per SQKM : 6km

浦东—上海科技馆
开发强度（容积率）0.8
综合建筑密度：18.2%
道路宽度：14~16米
每平方公里用地里的道路长度：8公里

Pudong - Shanghai science and technology museum
Development Intensity (FAR) : 0.8
Average Building Coverage : 18.2%
Width of Roads : 14~16m
Total Length of Road per SQKM : 8km

浦东—联洋社区
开放强度（容积率）2.3
综合建筑密度：15.5%
道路宽度：18~30米
每平方公里用地里的道路长度：4.8公里

Pudong - Lianyang community
Development Intensity (FAR) : 2.3
Average Building Coverage : 15.5%
Width of Roads : 18~30m
Total Length of Road per SQKM : 4.8km

浦东—金茂大厦，上海环球金融中心，上海中心
开发强度（容积率）：4.8
综合建筑密度：15.0%
道路宽度：24~30米
每平方公里用地里的道路长度：5.2公里

Pudong - Jinmao tower,The Shanghai world financial center,Shanghai center
Development Intensity (FAR) : 4.8
Average Building Coverage : 15.0%
Width of Roads : 24~30m
Total Length of Road per SQKM : 5.2km

上述特征揭示了步行网络主导下的城区和机动车干道主导下的城区在空间结构上的极大差异。不可否认的是，机动车干道主导的城市建设和发展已经成为一种客观趋势，得到广泛的接受和认同，它更符合当代人出行、生活的方式，并更好地服务于当代城市空间扩张的普遍需求。即便是从步行化时代就已经发展起来的上海浦西部分城区，它在机动车时代也必须经历机动化的改造，以符合上海新的发展规模和定位。上海的机动车道路结构目前已经基本成型，它由第一层级的高速干道（郊区环线、中环和由内环线、南北高架和延安路高架组成的"申"字形中心城高速路系统）和第二层级、第三层级的城市干道体系组成。这个体系构成了上海城市空间结构的主要骨架（图12）。北京的层层环路和网格形干道网络、深圳以深南路为中心延伸向两侧腹地的网格形干道，无不在书写着空间发展故事。以干道为骨架展开的城市结构，已经成为中国城市发展的基本范式。从另一个角度来看，有理由认为这个范式不仅是中国本土的，也是世界性的，在那些快速形成的新建城区和扩张型城市的发展中，该范式被广泛运用的例子屡见不鲜，无论是维也纳的多瑙新城、巴黎的德方斯城区，还是巴西的巴西利亚（新区）或是迪拜。

Above indexes and characteristics reveal the difference of between urban form dominated by pedestrian-friendly street net and that dominated by arterial traffic roads. However, the urban development led by arterial road has been widely accepted and acknowledged by contemporary cities. It meets the demand of contemporary people's transportation and inhabitation habit, while it also meets the demand of urban expansion which is the prevailing tendency of contemporary cities. Taking Shanghai again for example, even for those areas in Puxi which are famous for its high quality for inhabitation, they need to be restructured in a certain degree to give way for motor traffic(Fig.12). Similar stories take place in all cities in China. It must be also recognized that urban form dominated by arterial roads prevails not only in China, but is internationally adopted for rapidly developed areas such as Donau-City of Vienna, La De Fense of Paris, government district of Brazilia or Dubai.

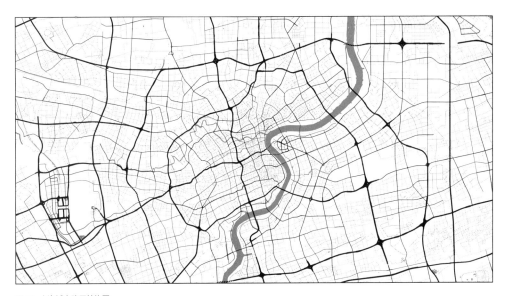

图12 上海城市路网结构图
Fig. 12 Road net of Shanghai today

3 "中心"还是"边缘"：
干道的命运转折
"CENTRAL" OR "PERIPHERAL": THE TURNING POINT OF ARTERIAL ROADS

　　随着城市发展的扩张阶段逐渐结束，对城市关注的角度也必须从"速度"转向"质量"。当人们不再热衷于日新月异的物质更新和基础设施改善，而转向关心城区里的生活便利程度、社会融合状况、城市活力、步行环境、公共空间和文化生活等话题的时候，干道主导的城市结构开始显示出它的局限。而干道周边的城区也在逐渐从城市发展的"中心"转向了"边缘"。

　　以上海为例，20世纪90年代以后规划的几处城市商业和金融聚集区均依托主要城市干道的交汇处设置，这充分显示了干道作为城市发展"中心地带"所带来的巨大支撑。例如徐家汇是城市西南部的中心区域，依托肇嘉浜路、虹桥路、漕溪北路、衡山路和华山路等五条主要道路；作为上海东北部中心区域的五角场依托的主要城市干道是翔殷路、邯郸路、四平路、淞沪路和黄兴路。依赖作为干道交汇点良好的机动车交通可达性和突出的视觉标志性，这些区域的发展初期都显示出强大的生命力；但随着建设规模的形成，这些地区的发展出现机动交通拥堵，户外环境恶劣，地块步行联系不便等一系列问题，成为这些区域进一步发展升级的阻碍。另外一些区域，建设初期依托单一的机动车道路，修建很多标志性建筑，却从来也没有成为受人欢迎的城区，例如延安路高架两侧和浦东陆家嘴世纪大道两侧的城区。反观一些并不依托机动车干道发展的城市区域，诸如原法租界所在的城区或古北的一些区域，却因为它们更适合步行交通而获得更高的空间质量，显示出可持续发展的良好趋势。

With the period of urban expansion coming to the end, the focus of urban development shall turn from "Speed" to "Quality". Living convenience, social intergration, urban vitality, safety, pedestrian-friendly urban environment and public space become new keywords for people. In this moment, urban structure dominated by arterial roads shows its limitations. And the urban areas alongside arterial roads, gradually turn from "center" of development to "periphery" of development.

In Shanghai, many commercial and financial areas had been planned and built after 1990s based on joining point of arterial road. Among them, Xujiahui acts as the central area of Shanghai's southwest region, which was built in the joining point of Zhaojiabang Road, Hongqiao Road, North Caoxi Road and Huashan Road. Wujiaochang, as central location of Northeast Region, was built at joining point of Xiangyin Road, Handan Road, Siping Road, Songhu Road and Huangxing Road. The location at joining point of arterial roads brings high accessibility and visual centrality, and thus strongly support these projects in their initial development period. However, with the projects and the surrounding area developed to certain stage, newly emerging problems such as traffic congestion, negative outdoor environment and lack of pedestrian linkage between buildings hinder the areas from further upgradation. In other projects, many landmark buildings were built simply alongside the arterial roads, to make use of outstanding visual centrality of the location. These areas never had become favorite places for people, such as those projects alongside Yan'an Road Elevated Road and Century Avenue in Lujiazui CBD. On the contrary, many other areas not being situated adjacent to arterial road, which might be treated by planning authority as lacking of accessibility, gain sustainable development power due to their pedestrian-friendly environment. In context of today's urban development of Shanghai, their location becomes more "central".

依托干道的城市发展方式遇到的问题在两个方面：其一是机动车的速度和噪声影响两侧步行活动的安全性和舒适度的问题。在这样的环境下，城市公共活动难以发生，也难以引起城市功能的进一步聚集。其二是机动车可达性在一定程度上成为步行可达的障碍，例如道路切割城市步行网络，导致地块之间的步行联系不便，妨碍城区空间作为系统整体的提升（图13）。然而，那些并非依托机动车干道发展的城区往往具有更好的塑造步行环境的条件，在城市公共交通极大提升的今天同样获得了很高的可达性，它们的发展呈现出很大的潜力——成为新的城市发展"中心"。

作为城市化前一阶段主要空间发展依托的机动车干道，在可以预见的未来是城市结构的重要组成部分。城市干道对步行活动是否友好，它的两侧是否能够组织起较高质量的城市空间，这些决定了它的将来是否能够持续作为城市发展的"中心"地带而存在。机动车干道两侧的步行环境提升必然成为我们这个时代应该正视和解决的问题。

Urban development alongside arterial road has to face up to two major challenges. The one is, the speed and noise of motor traffic make the areas alongside the road unsafe and uncomfortable places. Such unfriendly environment prevents the area from becoming an ideal place for public activities and functions. The second is, the motor traffic strongly hinder the accessibility by pedestrian. Arterial roads cut through and thus break down the pedestrian network, make the accessibility by pedestrian very low. The connection between building blocks on sides of arterial roads becomes very difficult(Fig.13). On the contrary, those other areas which are somehow away from arterial road have better possibility to form higher pedestrian quality. Today with highly developed public transportation, these areas gain also very high accessibility.

As a key urban element which had played such an important role in the previous stage of urbanization, arterial road will continue its spatial function at least in the near future. Whether pedestrian-friendly and high quality urban environment can be planned or added to its sides, determines whether the areas along it can continue to play the role as "central" place in city. The improvement of pedestrian-friendly environment alongside arterial roads shall be treated as one of the key problems in city that we shall face up and try to give a solution to in our era.

图13 被干道切割的城市环境 浦东陆家嘴
Fig. 13 Urban space separated by arterial roads,Lujiazui Pudong

4 城市沿干道发展的空间特征
SPATIAL CHARACTERISTICS OF AREAS ALONGSIDE ARTERIAL ROADS

机动车主干道为沿线用地提供较高的可达性，物流、人流、能量流、信息流得以流动和聚集，因而沿城市干道两侧用地比其他城市区域开发强度更高、发展也更迅速，这是城市沿路发展模式得以形成的主要基础。该模式主导下的城市结构特征与传统城市有很大不同，前文已经有论述。依托干道所发展起来的城区普遍具有如下空间特征：①干道两侧聚集大量高强度开发的地块，对干道的交通支持要求很高，视觉上往往形成强烈意象。②干道两侧的地块较大，支路网密度不高，与主干道交叉的接入口比较稀疏（规划上为了提高干道的平均车速，往往有意限制）。该特征导致干道对两侧支路网的可达性支持度不高。③干道两侧的建筑物退界较大。建筑物的出入口和消防很难由干道直接支持，必须在地块内形成一套自身的交通流线，再与两侧支路网连接，其后果是干道两侧人行道与建筑物再也无法建立关联。④由于建筑物建设主体各异，并需要满足规划设计对基地两侧的退界要求（如果要求建筑物贴线的城市设计导则不存在的话），干道两侧建筑界面往往不连续，难以为干道两侧人行道提供宜人街道尺度和功能支持，加之机动车造成的安全和噪声影响，干道两侧步行环境恶劣。⑤跨越干道步行交通比较困难，令干道成为两侧步行网络断裂的地带（图14—图17）。

For urban area developed alongside arterial roads, the following characteristics can be observed: 1) High-density areas agglomerate alongside arterial road, which demand also high traffic accessibility supported by arterial roads. These high-density areas always create strong visual images from the viewpoint of arterial road. 2) The blocks alongside arterial road are relatively larger while the street net is less dense. This tendency is more promoted by planning authority's attempt to reduce the number of joining point of neighboring streets to arterial roads to prevent lowering of the traffic speed on arterial road. 3) Building code requires big building set-back from the edge of arterial road, which detaches the building from direct contact to the road. The set-back distance sometimes is partly privatized by the properties on both sides of roads. To ensure the entrance to the building especially in fire emergency, an internal circulation has to be built within the area of building block. This further isolates building from the surrounding urban roads. 4) Building blocks on both sides with different development entities need to follow building code which demands on set-back from property lines in its two sides (if there is not an urban design guideline formulating the continuous street building surface). It results in fragmentary road surface which cannot provide human scale and proper urban functions for the sidewalks. With the situation being made worse by safety and noise problems brought by motor traffic, the sidewalk space alongside arterial road is not friendly for pedestrian. 5) It is hard to cross arterial road on ground level. This also makes arterial roads become an area where the street nets on both sides(if it does exist) break down (Fig. 14-17).

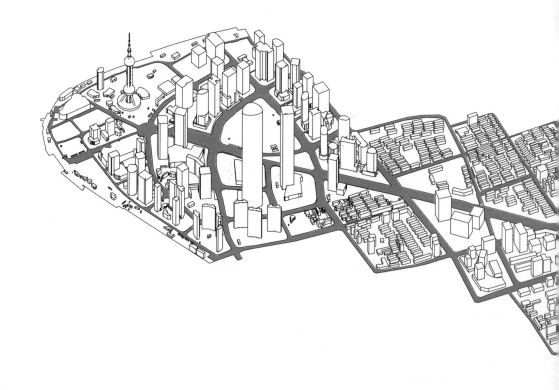

图14 世纪大道沿路发展的城区
Fig.14 Urban Areas alongside Century Avenue

上海浦东世纪大道规划宽度100米，从东方明珠至世纪公园全长约5.5公里。它是浦东陆家嘴CBD的空间主轴，串联东方明珠、陆家嘴公园、国金中心、金贸大厦、环球金融中心、上海中心，东至东方艺术中心和科技馆等重要城市地标建筑。世纪大道沿线是陆家嘴区域金融办公楼宇聚集的城区。

Being the spatial axis of Lujiazui CBD in Pudong Shanghai, the Century Avenue is 100 meters wide and 5.5 km long, linking the Oriental Pearl TV Tower, Lujiazui Park, IFC, Jinmao Tower, Global Financial Tower, Shanghai Center Tower, Oriental Art Center and the Museum of Technology. The areas alongside the Century Avenue are the places where high density development (mainly financial function) is situated.

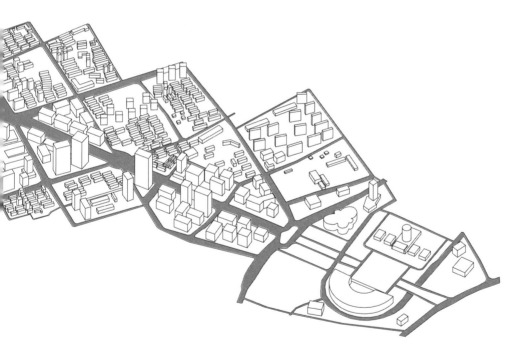

PEDESTRIAN PARADISE ON ARTERIAL ROAD

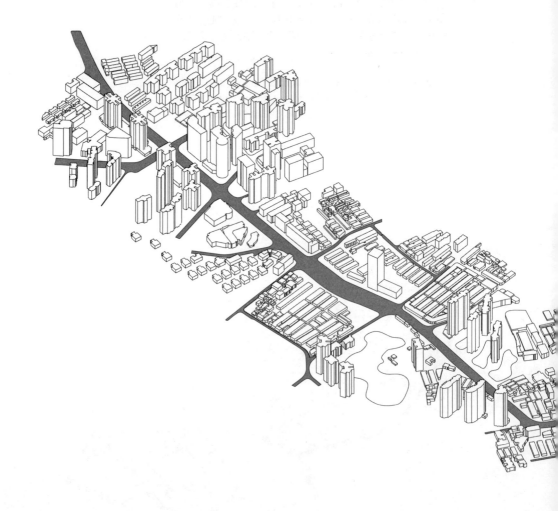

图15　四平沿路发展的城区
Fig.15　Urban Areas alongside Siping Road

上海四平路始建于20世纪30年代，是连接上海中心区域北部城区的主要干道，全长6公里，横跨杨浦、虹口两区。路北段连接五角场和新江湾城，是上海北部城市扩张的重点区域；路中段和南段混杂着很多历史城区，呈现城市开发和历史城区相互交融的状态。

Siping Road was initiated in 1930s, and since then has been acting as main linkage between central Shanghai and its northern region. With whole length of 6 km, it crosses over Yangpu District and Hongkou District. Its north end is connected with Wujiaochang and New Jiangwan Town, a major development area of northern Shanghai. In the middle and southern part of Siping Road many historic areas are situated, which is mixed with high-density urban development.

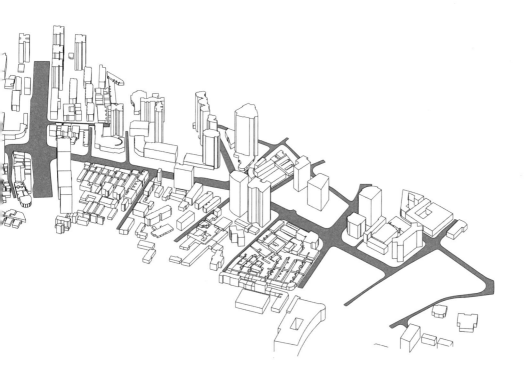

PEDESTRIAN PARADISE ON ARTERIAL ROAD

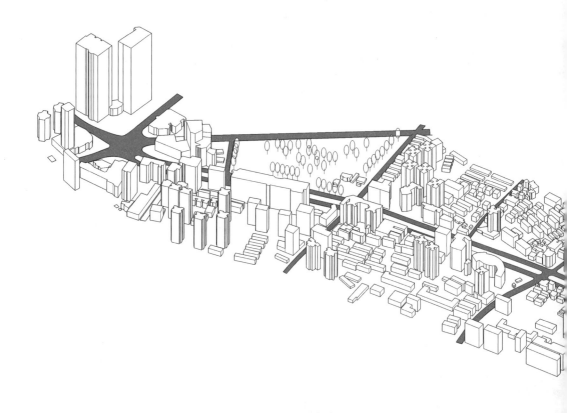

图16　肇嘉浜路沿路发展的城区
Fig.16 Urban Areas alongside Zhaojiabang Road

肇嘉浜原是上海地区一条东西走向的通航河流，20世纪50年代改造成道路，东起打浦桥，西至徐家汇，名为"肇嘉浜路"，全长4公里。肇嘉浜路两侧是高密度开发地带，两端的打浦桥和徐家汇是上海南区重要的商业中心区。

Zhaojiabang used to be part of Shanghai's waterway, and was planned and developed as arterial road since 1950s. Being 4 km long, it links the Dapuqiao in the east and Xujiahui in the west, both of which are important commercial areas in southern part of Shanghai. High density development takes place alongside the road.

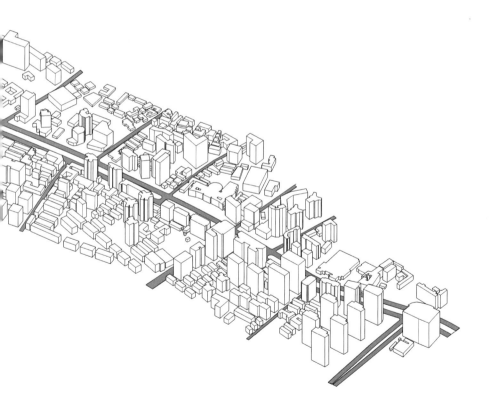

PEDESTRIAN PARADISE ON ARTERIAL ROAD

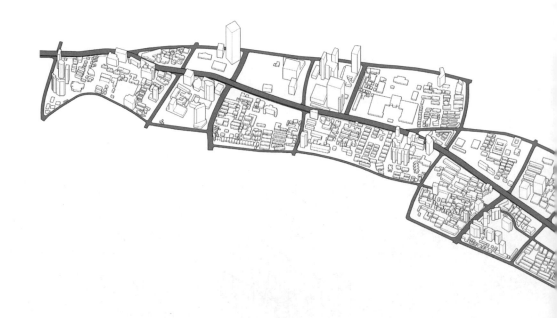

图17 延安路沿路发展的城区
Fig.17 Urban Areas alongside Yan'an Road

上海延安路西起虹桥机场，东至外滩，全长15公里。路中段与东段穿越上海市中心区（内环线以内的区域），其上架设延安高架路，成为双层机动车道。延安路是上海中心区的东西向空间主轴，沿线开发量巨大，也是上海城市形象的主要展示区域。

Yan'an Road is 15 km in total, which connects the Hongqiao Airport in the west and the Bund in the east. Its middle part and east part run across the central area of Shanghai (the area within the inner ring road). Elevated road was built after 1990s and thus Yan'an Road became a double-layered high-speed road. It acts as major spatial axis of Shanghai's inner city from the east to the west. The high-density urban development alongside the road makes it a highlighted viewing place for Shanghai's urban image.

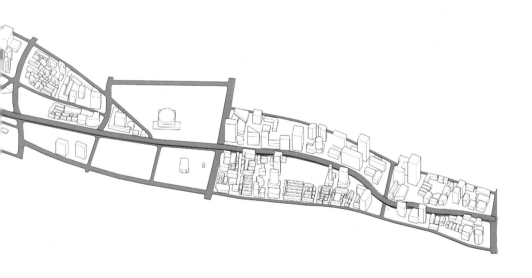

5 城市空间意象的变异

DUAL URBAN IMAGES IN MOTOR AGE

机动车时代的城市意象发生了巨大变异。与传统城市空间相比，城市意象出现了两种不同的体验角度：一种是驾车在机动车干道上体验的城市意象；另一种则是在人行道上体验的城市意象。从这两种视角上体验的城市意象有较大的差异，并且，随着道路宽度的增加，差异性也随之增大。首先，车行视角观察到的城市空间形象相对较为完整，反映了无论是城市规划还是建筑设计都给予该视角较高的重视度，而从人行视角观察到的形象则不够完整且碎片化；此外，人行视角的观察反映出人行道与建筑物之间缺乏联系，因而建筑物对人行道提供的功能支持较少，舒适的尺度感消失，城市活动无法展开。干道两侧的人行道完全成为干道的附属物，其功能退化为通行的交通设施，仅仅是从A点到B点的连线。街道作为城市公共空间的社会属性已荡然无存（图18—图21）。

Urban image during Motor Age have dual perspectives, the one is from pedestrian's viewpoint, so as in urban image in the traditional cities, while the other is from the viewpoint of motor vehicles. Urban images from the two viewpoints are very different, and the degree of difference becomes larger when the roads are wider. Investigation shows that urban images from the viewpoint of motor vehicle (center of the road) are much more integrative than those from the viewpoints of sidewalk. The later one is much more fragmentary, demonstrated by the breaking scales (people scale in contrast to the building scale), fragmentary and enclosed building surface and narrow sidewalk space which fails to provide place for urban activities. Pedestrian sidewalk becomes a sole transportation facility to link different locations, while its function as public space is ignored (Fig.18-21).

PEDESTRIAN PARADISE ON ARTERIAL ROAD

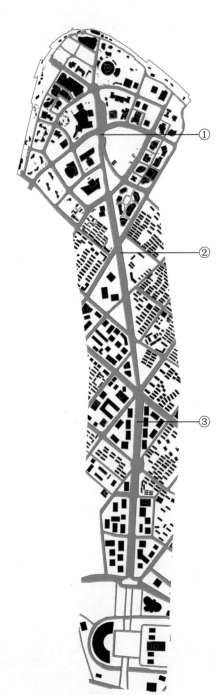

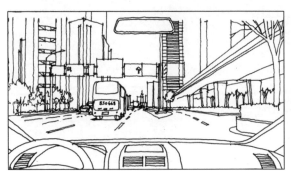

① 世纪大道—东泰路节点车行视角
Century Avenue -Dongtai Road
Perspective for Vehicle

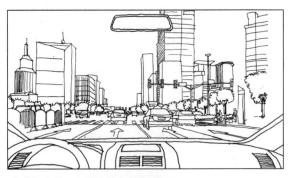

② 世纪大道—东昌路节点车行视角
Century Avenue -Dongchang Road
Perspective for Vehicle

③ 世纪大道地铁站节点车行视角
Century Avenue - Subway Station
Perspective for Vehicle

图18　世纪大道城市意象
Fig.18　Urban Images on Century Avenue

① 世纪大道—东泰路节点人行视角
Century Avenue -Dongtai Road
Perspective from the Sidewalks

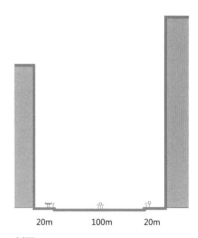

20m　100m　20m

剖面
Section

② 世纪大道—东昌路节点人行视角
Century Avenue -Dongchang Road
Perspective from the Sidewalks

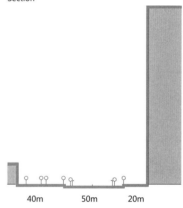

40m　50m　20m

剖面
Section

③ 世纪大道地铁站节点人行视角
Century Avenue -Subway Station
Perspective from the Sidewalks

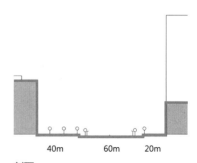

40m　60m　20m

剖面
Section

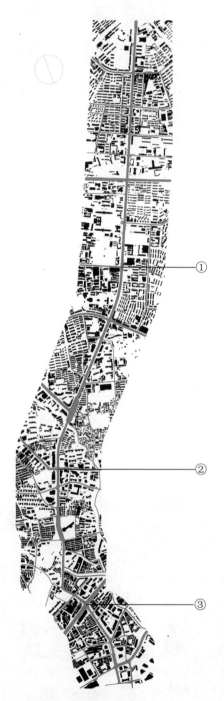

① 四平路—赤峰路节点车行视角
Siping Road -Chifeng Road
Perspective for Vehicle

② 四平路—临平路节点车行视角
Siping Road -Linping Road
Perspective for Vehicle

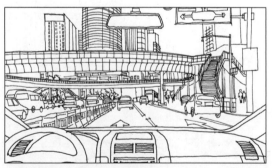

③ 四平路—海宁路节点车行视角
Siping Road -Haining Road
Perspective for Vehicle

图19　四平路城市意象
Fig.19　Urban Images on Siping road

① 四平路—赤峰路节点人行视角
Siping Road -Chifeng Road
Perspective from the Sidewalks

12m　25m　10m

剖面
Section

② 四平路—临平路节点人行视角
Siping Road -Linping Road
Perspective from the Sidewalks

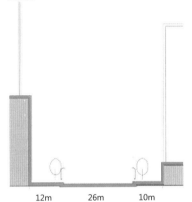

12m　26m　10m

剖面
Section

③ 四平路—海宁路节点人行视角
Siping Road -Haining Road
Perspective from the Sidewalks

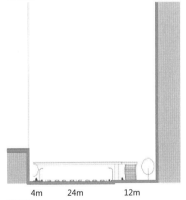

4m　24m　12m

剖面
Section

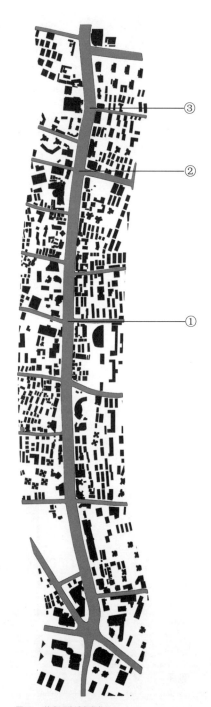

① 肇嘉浜路—枫林路节点车行视角
Zhaojiabang Road -Fenglin Road
Perspective for Vehicle

② 肇嘉浜路—大木桥路节点车行视角
Zhaojiabang Road -Damuqiao Road
Perspective for Vehicle

③ 肇嘉浜路—雅安路节点车行视角
Zhaojiabang Road -Yan'an Road
Perspective for Vehicle

图20　肇嘉浜路城市意象
Fig. 20 Urban Images on Zhaojiabang Road

① 肇嘉浜路—枫林路节点人行视角
Zhaojiabang Road -Fenglin Road
Perspective from the Sidewalks

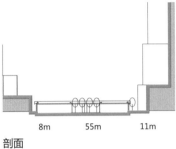

8m　55m　11m

剖面
Section

② 肇嘉浜路—大木桥路节点人行视角
Zhaojiabang Road -Damuqiao Road
Perspective from the Sidewalks

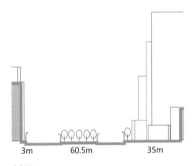

3m　60.5m　35m

剖面
Section

③ 肇嘉浜路—雅安路节点人行视角
Zhaojiabang Road -Yan'an Road
Perspective from the Sidewalks

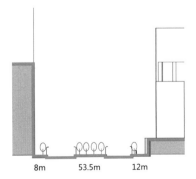

8m　53.5m　12m

剖面
Section

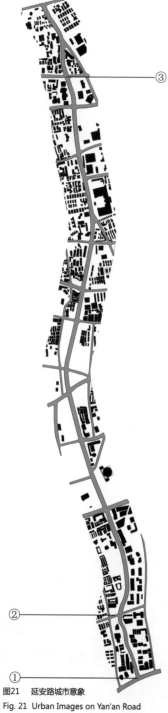

① 延安路—中山东一路节点车行视角
Yan'an Road -Zhongshan North No.1 Road
Perspective for Vehicle

② 延安路—河南中路节点车行视角
Yan'an Road -Henan Middle Road
Perspective for Vehicle

③ 延安路—华山路节点车行视角
Yan'an Road -Huashan Road
Perspective for Vehicle

图21　延安路城市意象
Fig. 21 Urban Images on Yan'an Road

① 延安路—中山东一路节点人行视角
Yan'an Road -Zhongshan North No.1 Road
Perspective for Sidewalk

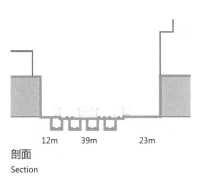

12m　39m　23m

剖面
Section

② 延安路—河南中路节点人行视角
Yan'an Road -Henan Middle Road
Perspective for Sidewalk

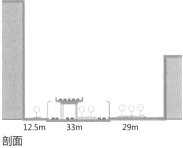

12.5m　33m　29m

剖面
Section

③ 延安路—华山路节点人行视角
Yan'an Road -Huashan Road
Perspective for Sidewalk

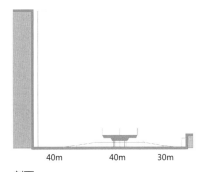

40m　40m　30m

剖面
Section

6 城市空间可步行性研究的基础和进展
RESEARCH BACKGROUNDS AND PROGRESS ON URBAN WALKABILITY

机动车时代城市步行环境的恶化使"可步行性"一词成为城市设计领域研究的热议话题。早在20世纪60年代，简·雅各布斯就在其著作《美国大城市的生与死》中指出"城市的本质就是为了人们的生活"，倡导"适宜的尺度"和"宜居社区"。她认为：城市空间应该具有多样性和复杂性；一个好的城市应该在街道尺度上易于开展社会交往；街道和公共空间必须是可步行（walkable）的（J.Jacobs，1961）。其著作对当代城市规划的学术研究和实践产生了深远的影响。

迈克尔·索斯沃斯认为良好的步行环境是鼓励低碳出行的必要条件。可步行性是建成环境支持和鼓励行走的程度，其衡量指标包括为行人提供舒适安全的环境，在合理的时间和成本内使人们能够到达各种目的地，并在步行道网络中提供行程上的视觉吸引（M.Southworth,2005）。从这个定义中可以看出可步行性的影响因素诸多，如公共空间网络的便捷度、公共空间网络的品质和安全以及服务设施的分布等。建成环境对步行的支持和友好程度主要可以从两个方面来体现：一，建成环境的特征；二，步行出行的特征。从第一点来看，影响建成环境的因素主要包括：土地开发模式、道路网络特征及公共空间网络环境品质。从第二点来看，影响步行出行的因素主要包括：居民的出行目的、出行距离、出行次数。在此基础上索斯沃斯提出了成功步行系统的六条评价标准：①连续性；②与其他系统的连接性；③精细化土地利用模式；④安全性；⑤路径的品质；⑥路径的环境。

关于可步行性的研究，在城市设计、城市交通、城市安全、健康与场所、社会学等领域有很多学者进行了各种类型的研究。这些研究主要集中在三个方面：一是关于可步行性的计量方式的研究；二是关于城市空间是否适宜于步行活动的评估；三是探索一揽子解决城市步行空间问题的方法。安·福赛思对所有关于可步行性研究文献进行梳理，总结了九个方面：①步行出行的方便性；②目的地（设施）的步行距离；③对步行行为的支持；④安全性；⑤宜居与社交；⑥体育锻炼；⑦可持续交通选择；⑧多维度；⑨整体解决方案（A.Forsyth,2015）。

How to deal with the deterioration of pedestrian environment in the Motor Age has become an important topic for urban researchers, among which "walkability" is one of the keywords. As early as in 1960s, Jane Jacobs has pointed out in her *The Death and Life of Great American Cities* "the nature of city that it is a place for people's life" and the importance of urban street as place for social communication and interaction. Street and public space shall both be walkable (J Jacobs, 1961). This book has profound influence on urban research and practice.

Michael Southworth points out that pedestrian environment is a basic condition for low-carbon city. Walkability reflects how well the built-up environment supports and encourages pedestrian behavior (M Southworth, 2005). Indicators for walkability include comfortable and safe environment provided for pedestrian and the possibility for pedestrian to cover multiple destinations within certain time and cost. Characteristic of built-up environment and the characteristic of travel are the two factors to determine walkability. The former includes land use model, street net, public space net and their quality. The second one includes inhabitants' travel destination, distance and frequency. Based on this, Southworth puts forward six criteria for pedestrian system: 1) connectivity; 2) linkage with other modes; 3) fine grained land use patterns; 4) safety; 5) quality of path; and 6) path context.

Many scholars research on walkability from varied disciplines of urban design, urban transportation, urban safety, health and place as well as sociology. These research can differentiate in directions, the first is on the quantification method of walkability, the second is on evaluation of urban space in its qualification to be walkable, and the third is on the design method to improve walkability of urban space. Ann Forsyth summarizes 9 domains in research of walkability: 1)traversable; 2)compact; 3)physically enticing; 4)safety; 5)lively and sociable; 6)exercise-inducing environments; 7)sustainable transportation options; 8)multidimensional; 9)holistic solution(Ann Forsyth, 2015).

在可步行性计量方式方面，目前主要有步行审计（walk audit）、步行指数（walk score）等方式。可步行性主要表达了建成环境如何去实现步行的目的，它与步行者行为与所使用的交通工具（Frank & Engelke, 2005; Owen, Humpel, Leslie, Bauman & Sallis, 2004; Sallis, Frank, Saelens & Kraft, 2004）、用地功能设置（包括生活、工作、休闲等）、路径的方向及不同目的地之间的联系有关（Forsyth & Southworth, 2008; Moudon et al., 2006）。步行指数是判断区域步行友好程度的指标。人们可以在步行指数网站的地图上找到步行距离内的商店、餐馆、酒吧、公园和其他设施的地址，使步行出行的意愿得以提高。该网站的服务目前已覆盖美国、加拿大和英国的各大城市。步行指数的算法主要采用了从居住者到各类设施的步行距离以及在步行距离内所覆盖的各类设施的数量。邓肯等研究了步行指数在美国大城市各种空间尺度上的评估能力，认为步行指数与社区的可步行性有很大相关性（Duncan,2013）。

在关于城市空间是否适宜于步行活动的评估方面，可步行性主要关注社区的宜居性以及公共空间是否利于进行社交活动和体育锻炼等。在这类的研究中，学者们通常认为可步行性与城市活力和社区生活等要素相关联，例如，威廉·怀特（W.Whyte,1980）、杨·盖尔（J.Gehl, 2011）等学者从社会学角度探讨如何设计适宜步行活动的公共空间(诸如社区商业、功能混合、密度等)，而步行活动本身并不是研究的对象。杰夫·史贝克（J.Speck）在其著作《适宜步行的城市》（Walkable City）中指出，可步行性既是目标也是工具，可步行性也许是对城市空间活力最有帮助的要素，一旦提升了城市的可步行性，城市的宜居性和活力便随之而来（J.Speck,2013）。史贝克在书中分别从城市步行系统的实用性、安全性、舒适性、趣味性四个方面详细介绍如何采用十个步骤创建可行走的城市，提高城市的步行指数。此外，也有大量研究是关于可步行城市空间如何通过促进体育锻炼以创建健康城市。在这个语境下，研究主要关注步行行为作为身体的活动。杉山（Sugiyama,2012）发现大多数研究关注于步行出行与目的地或用地功能之间的关系，提出应该规划更为连续的路径和适宜运动的步行环境。盖博（Gebel）认为具可步行性的场所一般都拥有比较方便进行体育锻炼的环境(Gebel et al., 2009)。

在探索解决城市步行环境问题的研究中，可步行性是一个衡量良好城市环境的总称，正如艾米丽·泰伦（E.Talen）和朱丽亚·科希斯基（J.Koschinsky,2013）在其对于可步行社区的文献分析中所说"一个场所或一个社区应该具有可步行性的想法早在19世纪早期就已经出现，是一种关于健康、和谐、市民化、文明生活、民主、精神健康、社会公平等概念的全面的概括"。他们认为可步行性是通向和谐城市的基石，是可持续城市的基础（无论是环境可持续还

In terms of research about quantification method, "walk audit" and "work score" are the two main indicators to quantify Walkability. Walk Score is an indicator to reflect how friendly the environment is to the pedestrian behavior. In the website of Walk Score, people can identify shops, restaurants, bars, parks and other facilities in his walking distance, and thus their desire to walk is enhanced. The service of this website can cover USA, Canada and major cities in GB. The algorithm of walk score is mainly based on the walking distance of people to facilities and the number of facilities within the walking distance. Duncan's research (2013) reveals the strong relevance between Walk Score and walkability of the urban area, based on analysis for big cities in USA.

In terms of research domain on evaluation of urban space with its qualification to be walkable, walkability focuses on whether the built-up environment is suitable for living, social communication and sport. Scholars tend to associate walkability with urban vitality and urban life. Researches made by William Whyte (William Whyte,1980) and Jan Gehl (Jan Gehl, 2011) both focus on how pedestrian-friendly urban space can be designed (such as community shops, function mix and density) from perspectives of sociology, while the behavior of waking is not an object of research. Jeff Speck points out in his *Walkable City* that the behavior of walking is an important content of urban life. Once a city is walkable, the urban vitality is naturally generated(Jeff Speck,2013). Speck further puts forward 10 steps to establish walkable city from 4 aspects of urban pedestrian system, which is the useful walk, the safe walk, the confortable walk and the interesting walk. On the other hand, many other researches focus on how walkable city is related with inhabitants' health by promoting sport. In this context, walking itself is treated as one kind of sport. Sugiyama (2012) points out that most researches take important attention on relation between walking and destination as well as land use, and proposes to plan continuous pedestrian-friendly path and environment suitable for sport. Gebel points out that, walkable places are normally a place better suitable for sport (Gebel, 2009).

In terms of research domain on method to improve walkability of urban space, Talen and Koschinsky point out that, based on literature research on walkable communities, the thought that a place or community must be walkable rose in early 19th century, which is a demonstration of concepts about health, harmony, civilization, urban life, democracy and social equality. Walkability is a basis for sustainability in aspects of environment, economy and sociology, as well as a ladder leading to harmonious city. In such context, walkability becomes a key element to

是经济、社会可持续），也是促进社会交往、减少犯罪和减少其他社会问题的根本。可步行性在这个语境下通常成为一个吸引投资进行城市再开发、吸引居住人口和提高宜居性的关键指标（Addison et al., 2013）。

综上所述，在机动车主导的当代城市环境中，可步行性问题是衡量一个城市是否宜居、是否可持续的一个重要指标。随着机动车时代的干道成为城市发展的基本要素，城市发展区域向干道周边转移，当代城市的步行环境大多数是在城市干道的两侧开展。因此，将目光聚焦于干道，解决干道上的步行问题将是一个非常重要的研究领域，其研究成果的应用也将为当代城市的可步行性带来较大改观。

attract investment for urban redevelopment, to attract inhabitants and to improve livability of urban area (Addison, 2013).

In conclusion, in the Motor Age, walkability is an important indicator determining whether a city is livable and sustainable. Since arterial road has become dominating element in contemporary cities and many urban areas have been developed alongside arterial roads, sidewalk space of arterial roads becomes very important place (or majority of place in terms of amount) in city for pedestrian behavior. How to improve the pedestrian quality alongside arterial roads therefore becomes an important research domain today. The implementation of research may have significant influence on contemporary city's built up environment.

7 干道与步行的合集：
愿景与行动
PEDESTRIAN PARADISE ON ARTERIAL ROAD: VISION AND ACTION

前面的论述导向一个结论：高机动车可达性固然是城市发展最重要的条件之一，但创造适宜步行的城市区域才是城市发展最终的目的，因为良好的步行环境决定了一个城区作为工作、休闲和居住城区的基本环境质量。城市干道对步行活动是否友好、它的两侧是否能够组织起适合步行的城市空间决定了城市干道将来是否能够持续作为城市发展的"中心"地带存在。从城市空间结构的层面上看，既然在当代城市中城市干道已经成为城市空间结构的骨架，提升它的可步行性无疑意味着整个系统可步行性的极大提高。这就是我们提出"干道与步行的合集"这样理念的理性基础。

另一个方面，既然现阶段城市发展的重点已经从"空间扩张"转向"质量提升"，从交通发展角度也必然从干道建设转向公共交通建设。很多发达国家城市中心区的出行已经转向公共交通和以骑行为主、个人小汽车为辅的模式，城市干道已经日益失去了其功能，对其进行骑行与步行化改造具有天然的条件。

例如维也纳的指环路，其作为老城边缘的第一层环道，曾经是机动车交通的干道。一百年来的陆续改造和提升，通过合理的断面设置（如限制宽度的主要道路、容纳减速交通的丰富辅道、电车轨道、自行车道和宽阔且优美的步行区域）、交通节点设置（单向转弯的车流、地下步行路径的穿越和连接）以及公共空间节点的设置，指环路保持着极高的环境质量和城市活力状态。20世纪80年代以后指环路上又陆续增加设置地铁站点，极大地降低了机动车的流量（图22）。维也纳另一条主要道路玛利亚大街（Mariahilferstrasse）曾经作为城市中心区通往西区的主要道路，道路宽度设计为四车道，两侧人行道仅为3米左右，步行环境不佳。20世纪初期经过改造缩减为两车道加两侧非机动车道和部分停车位，人行道扩大到5~8米，在起始点和中段增加三个地铁站点，从而使步行环境得到初步改善，引发两侧商业繁荣。2015年进行第二次改造，运用混合交通的概念，在大部分区域只允许步行、公交车和自行车进入，

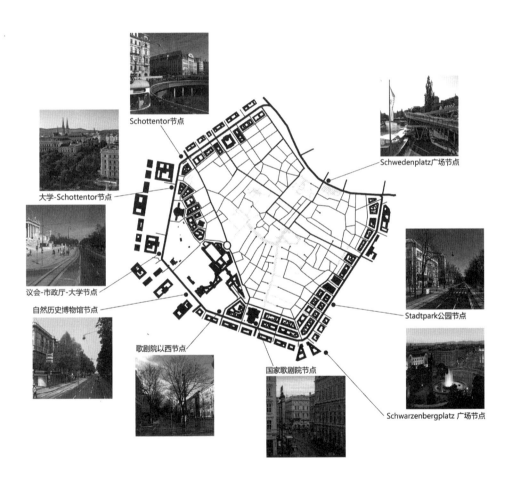

图22 维也纳指环路及其城市节点与界面
Fig.22 Ringstrasse and its urban nodes and street surface

小部分区域仍允许单向机动车进入，增加城市家具和场地铺装，令玛利亚大街的步行环境进一步提升，成为非常宜人的公共空间（图23）。

另一个例子是韩国首尔的清溪川。它曾经是首尔水系的一部分，从20世纪50年代起被填埋，20世纪70年代成为城市主要机动车干道，有两段甚至变成高架道路，这里曾经是首尔交通堵塞最严重的地方，道路两侧也成为环境非常恶劣的城市区域。2005年，经过重新规划，高架道路被拆除并恢复原水系，道路被缩减为两车道，分别位于河道的两侧，沿河道规划生态廊道与公共空间。该规划令清溪川成为非常宜人的城市区域，吸引很多市民与参观者，清溪川两侧成为高档住宅、城市商业的聚集地（图24）。

The aforementioned argument leads to a conclusion, which is that, although high traffic accessibility is one of the important conditions for the urban development, the establishment of pedestrian-friendly environment is the eventual target of urban development, because it determines the urban area's quality for working, leisure and habitation. Whether the urban area alongside arterial roads is pedestrian-friendly, determines whether the area can continue its role as "central" area of urban development. In term of spatial structure of the city, since arterial roads have become spatial aortas of modern city's structure, the improvement of its pedestrian quality will surely bring the improvement of walkability of the entire city. This is the rational basis of the concept of "pedestrian and arterial road hybrid" in this book.

On the other hand, since the current tendency of urban development is turned form "spatial expansion" to "quality improvement", the focus of urban transportation shall also turn from planning of arterial road to that of public transportation. In many cities in developed countries, the majority of urban transportation is public transportation and riding, while the personal automobile takes no more the majority. In such context, urban arterial roads lost its functions of traffic transportation, and urban design to improvement their pedestrian quality gains basic condition. One example is the Ringstrasse in Vienna, which acts as the important arterial road of Vienna. For 100 years, constant upgradation has been made to the road through section design (with limitation of traffic road, added affiliated road to lower traffic speed, trams way rails, bicycle lane and beautiful pedestrian sidewalks) and urban design on urban node (such added underground links and commercial passges, added plazas and parks). After 1980s, several new metro lines had been added to further release the road from traffic. The Ringstrasse remains its role as most beautiful urban space in Vienna (Fig.22).

The other important road in Vienna, Mariahilferstrasse, uses to be the arterial traffic road to link the core area of Vienna to its western peripheral area. The width of road is limited by traditional buildings. It had been planned with 4 traffic lanes with sidewalks only around 3 meters wide. Since 1990s it was planned newly with 2 traffic lanes with sidewalks being enlarged to 5-8 meters, while 3 metro stations were established in both ends and its middle part. This effort triggered prospering of commercial activities on both sides ever since. The second reconstruction took place in 2015, in which a shared street concept is introduced. Majority part of road is changed in to area in which only pedestrian, bicycle and bus are allowed, while the rest part being turned for single-lane traffic with very wide sidewalk. Urban space on the road is planned with landscape, hard pavement and urban furniture (Fig.23).

Another example is Cheonggyecheon in Seoul of Korea. It uses to be part of the waterway of Seoul, which was buried in 1950s and arterial roads were built on it in 1970s, with parts being occupied by elevated roads. It was the place perplexed by very severe traffic jam and big environment problem led to urban decay in its surrounding areas. After 2005, a new plan was made, in which the elevated roads and traffic roads under them were demolished and the waterbody was restored. Alongside the water ,ecological corridor and public spaces were established. Today the area alongside it becomes very livable place which attracts many citizens and visitors. High quality residential buildings and commercial facilities start to agglomerate in this area (Fig.24).

PEDESTRIAN PARADISE ON ARTERIAL ROAD

图23 维也纳玛利亚大街改造工程
Fig. 23 Renovation of Mariahilferstrasse in Vienna

1970—2000

今天
Today

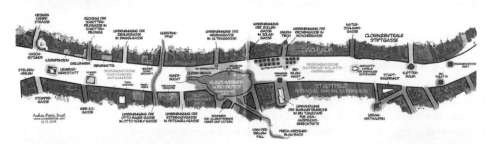

图24 韩国首尔清溪川改造工程
Fig.24 Renovation of Cheonggyecheon in Seuol

1973—2005

今天
Today

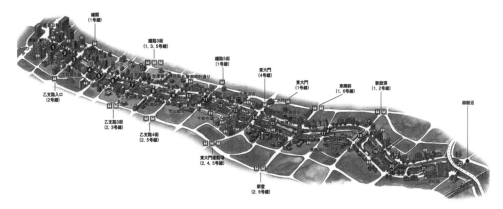

从城市设计角度，干道步行化优化的突出挑战在解决机动车交通和两侧城区步行环境的矛盾，其重点在于重新审视功能分区规划理念中的将各种城市要素分离的做法，使所有相关的城市要素再次组合。与干道可步行性相关的城市要素有：人行道及其边界界面、相邻的城市公共空间、相邻的建筑物、地下空间、绿化空间等等。本研究认为，上述这些要素需要在"点""线""面"三个层次上进行组合，以达到步行与干道耦合的目的（图25）。

From perspective of urban design, the challenge of pedestrian improvement of arterial road lies in how to solve the contradiction between traffic roads and pedestrian-friendly places on both sides. Focus shall be put on the reorganization and integration of urban element, which use to be separated by routine urban planning. Those elements include sidewalk and its urban surface, interconnected public space, adjacent buildings, underground spaces and greeneries. The elements shall be reorganized and integrated in the following 3 layers:dot, line, face, to achieve the goal of walking and artery coupling (Fig.25).

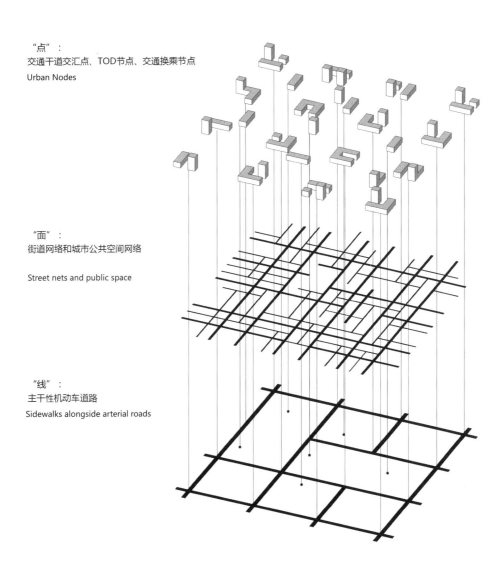

图25 干道步行空间优化的三个级别层面
Fig.25 3 layers of urban design on pedestrain space alongside arterial road

一、"点"层次上的耦合

"点"即城市干道的交叉点,是城市环境内最为活跃的部分。由于干道交汇带来的机动车可达性最高,人气最为旺盛,同时人车矛盾也最为突出,一般解决方案是架设人行天桥、地道,但由于不符合行人的行为特点,通过行人绕远来解决问题只能是权宜之计。本研究提出,必须积极鼓励临近建筑物参与到步行系统中,为天桥或地道提供行人缓冲空间,同时在行人路径上提供功能支持、行人设施(如室外电梯、自动扶梯,绿化、休息座椅等)和良好的城市公共空间,使行人穿越干道不再是一种单一的行为,使步行行为的可选择性提高,使干道交叉口成为立体"公园"(图26)。

1. Urban nodes

They are situated normally in the joining points of arterial roads, acting as most active parts in the built-up environment due to high traffic accessibility. These are also places where the most severe conflicts between traffic and pedestrian exist. Adding elevated bridges and underground corridors helping pedestrian to cross street is only a function solution. A more integrative system composed of adjacent building with public functions (shops and restaurants), sidewalks, greeneries, small squares, facilities for pedestrians (such as public elevators, escalators and urban furniture), together with elevated bridges and corridors shall be established to provide full function and urban use for passengers. Thus the crossing of road is turned from functional activity to urban experience. The joining point of arterial roads shall become 3-dimentional park for people (Fig.26).

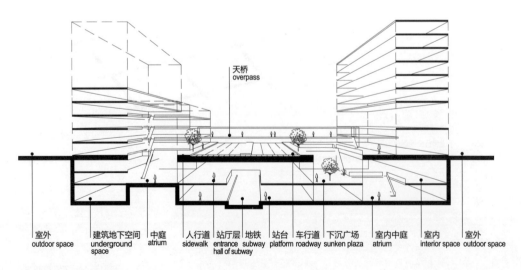

图26 "点"层次上的耦合
Fig.26 Urban Nodes

二、"线"层次上的耦合

对于城市干道而言,大量步行空间依旧是线性的人行道,因而线性路径步行环境的品质决定了整个体系的可步行性。由于当代城市空间格局的特点,干道交叉口距离通常在300~500米,在这个距离内如果城市用地恰好是封闭小区、单位(如大学、中学等),那么在300~500米的人行路径上,其边界很可能是围墙,使得行走变得索然无味,这是一个非常普遍的问题。即便不是围墙,通常也是与人行空间无法产生关联的绿化、停车位等等。本研究提出,需要建立步行路径与对步行活动友好界面的耦合关系。这种耦合关系建立在两者的贴合关系以及界面对于行人活动的支持程度。后者如人行路径的边界是否具有商业功能,是否提供扩展空间,是否提供遮风避雨设施,是否具有供行人休息、停留的空间,是否有座椅以及其他行人设施(如灯光、绿化、书报亭等)等等。一旦满足这些条件,即可认为路径与界面实现了一定程度的耦合,而耦合程度的高低则可根据具体情况进行设计(图27)。

2. Sidewalks alongside arterial road

Sidewalk alongside arterial roads takes the majority amount of pedestrian space in city, that is why its quality of environment will greatly affect the performance of the pedestrian of city as a whole. In new-built urban areas, the distance between joining points of traffic roads normally takes 300 to 500 meters, which is much wider than that in the traditional urban structure. The block's boundary alongside the sidewalk therefore is very important. If the boundary is composed of walls, fences, isolation greeneries or parking spaces, the pedestrian environment in this distance will become very unpleasant for passengers. The enhancement of environment quality lies in the relation between sidewalk and pedestrian-friendly urban surface (as boundary), which is jointly determined by how the two elements close to each other and how is the quality of urban surface. The later one is a result of composite factors such whether the surface provide commercial functions to sidewalk, expanded urban spaces, shelter spaces for rain and wind, leisure space, facilities (sitting, lights, greeneries, booths) etc. (Fig.27).

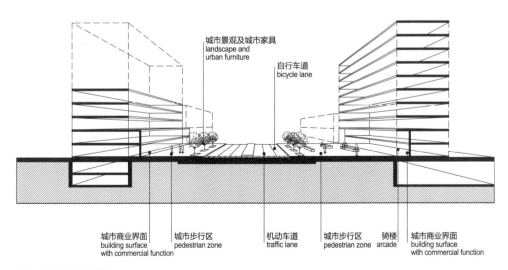

图27 "线"层次上的耦合
Fig.27 Sidewalks of arterial road

三、"面"层次上的耦合

城市的发展应该从沿干道的发展区域进一步向垂直干道的纵深方向城区扩展，以形成区域的整体发展和提升。这需要沿干道两侧建立适合步行的城市街道网络，并增加它们与干道的连接。此外，干道不应该成为街道网络断裂的地方。街道网络将通过步行环境提升的干道进行连接，形成跨路、跨区域的大型连续步行网络，推动城区作为整体的环境提升。该连接同样应该通过多层面步行系统实现，需要对建筑、景观、公共空间、服务设施进行多要素整合设计，而以此形成的跨路、跨区域超级步行网络则有可能成为机动车时代下一个阶段全新的城市空间类型（图28）。

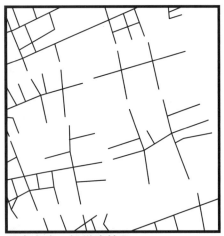

河南中路—福州路某段
Henan middle Road-Fuzhou Road

河南中路—延安路某段
Henan middle Road-Yan'an Road

西藏南路—延安中路某段
Xizang south Road-Yan'an middle Road

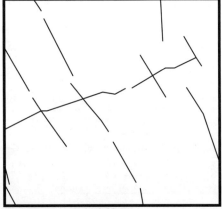

南京西路—镇宁路某段
Nanjing west Road-Zhenning Road

图28 "面"层次上的耦合
Fig. 28 Street nets and its integration through arterial road

3. Street nets and its integration through arterial road

The next stage of development of city shall grow from areas alongside arterial road to areas in its perpendicular direction, thus achieving integrative urban development in broader scope. For this, densely laid and pedestrian-friendly urban street net shall be planned in urban areas on both sides of arterial road and their connection with the arterial road shall be enhanced. Secondly, the arterial road shall not be place where the street nets break down. Instead, through carefully urban design that integrates building, public space, landscape and facilities, street net will be linked together. As a vision for future, continuous and cross-road pedestrian-friendly street net will be the new urban element to be generated in the Motor Age (Fig.28).

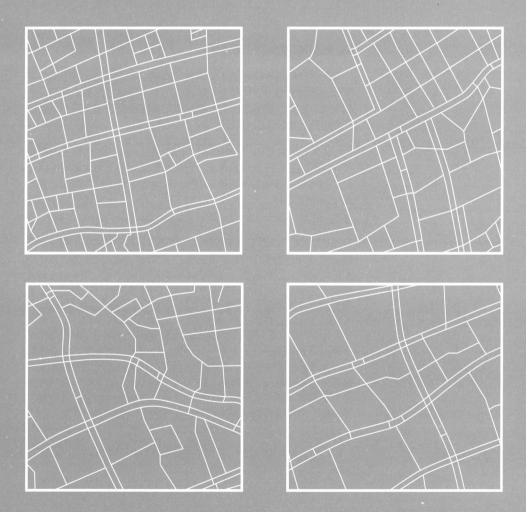

8 城市设计的实践：
四平路沿线步行系统优化
URBAN DESIGN PRACTICE：
TO IMPROVE PEDESTRIAN ENVIRONMENT ALONGSIDE SIPING ROAD OF SHANGHAI

■ **四平路作为城市干道的历史**

四平路辟筑于民国初年，原名"其美路"，以纪念曾经担任沪军都督的同盟会元老陈英士（字其美），其道路线形基本通直、大致呈南北走向。四平路作为一条城市干道曾经先后起到三个重要的联系作用，它们分别是：

（1）20世纪初，吴淞方向上的沪北江湾华界地区与上海租界市中心的连接。这种联系体现了近代城市化中的上海大都市区结构性扩张与内部加强整合的基本态势。江湾地区随后成为民国上海特别市政府推动的"大上海都市计划"的选定区，五角场的城市地理概念亦于此中发生（图29—图31）。

（2）淞沪战后，出于战略地位和现实条件，日本在上海重点经营吴淞—江湾地区以实巩固军政和经济领地，相应提出了以杉江房造编制的沪北"新市街"城市总体规划和发展计划。四平路在其时担当了这一新的日据地和国际租界北四川路吴淞路一带旧日据地的交通联系。四平路上一度辟筑了小型火车线路，供快速交通之用。

（3）解放后，四平路成为包含五角场的杨浦区北部新兴城市化区域与上海城市中心的主要联系通道。在该区域集中的大量高等院校、军事机构和大规模、多类型的居住小区迄今依靠四平路和吴淞路与外滩直接相连。相应的，在20世纪80年代以后的新一轮城市快速大发展中，四平路除作为交通轴之外，还成为中心城北部重要的发展轴，沿线城市更新活动十分活跃（图32，图33）。

图29 20世纪30年代大上海计划中的五角场中心区
Fig. 29 Wujiaochang in "The Greater Shanghai Plan" in 1930s

图30 20世纪30年代大上海计划中的其美路
Fig. 30 Qimei Road in "The Greater Shanghai Plan" in 1930s

Siping Road's History as Arterial Road

Siping Road was planned and developed since beginning of the Republic of China (1921—1949 in mainland China). It was named as "Qimei Road" before 1949 to pay tribute to senior founding member of Tung Meng Hui (Chinese Revolutionary League). As an important arterial road, it took its function as regional linkage in three periods of Shanghai urban development:

(1) In beginning of 20th century, it served as the important and only linkage between North-Shanghai Jiangwan Chinese settlement area and the central area of Shanghai which was international settlement. It represents the structural spatial expansion and internal integration of Shanghai as a metropolitan city in beginning of 20th century. Jiangwan Area later became the selected area to be developed as center of "Shanghai metropolitan plan", and the area called Wujiaochang, which is situated in the northern end of Siping Road, gained it ground in this period(Fig.29-31).

图31 20世纪30年代的五角场
Fig. 31 Wujiaochang in 1930s

(2) After the Songhu Battle in 1937, Japan occupied northern part of Shanghai and planned developed this area as its military, political and economic center in Shanghai. Siping Road functioned during this time as linkage of the Japanese-occupied area to the Japanese immigrant settlement area in North Sichuan Road - Wusong Road area in central part of Shanghai. Tramway was built in this period to provide rapid commuter transportation between two urban regions.

(3) After the foundation of People's Republic of China in 1949, Siping functions as linkage between northern part of Yangpu District, which is a newly developed urbanization area, and the central part of Shanghai. The many universities, military institutions and big-scaled residential areas developed in its northern part today still depend very much on it for connection to central part of Shanghai. In correspondence, heated urban development after 1980s had been taking the area alongside Siping Road as important development axis(Fig.32,33).

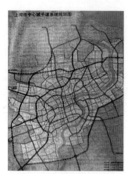

图32 1986年总体规划干道路网图
Fig. 32 Aeterial road net in Shanghai Master Plan 1986

图33 1999年总体规划干道路网图
Fig. 33 Aeterial road net in Shanghai Master Plan 1999

■ 四平路的沿线城市发展历程与特征

今天，四平路作为一条跨越杨浦、虹口两个城市行政区、全长超过6公里的城市主干道，随着不同的城市化发展进程而产生了类型多样的沿线城市空间，自北向南大致可分为三段来认识其主要特征：

(1) 北段：五角场—中山北二路

该段道路沿线曾经主要由空军政治学院、空军第四军军部等大型军事机构的"大院"和四平村等农业地带所构成。北段在20世纪80年代以来经历的城市形态变化十分显著，首先是1983—1984年第五届全运会的拓路工程，使四平路成为一条宽阔而美丽的景观大道；大批八九十年代的住宅小区建设、临近五角场的城市商业设施开发以及1994年中山北二路高架道路的建成进一步彻底改变了曾经的农田与大院相间的城市景观。

(2) 中段：中山北二路—溧阳路

该段沿线的城市化在20世纪80年代之前已经完成，始建于50年代的沪东工人居住区——鞍山新村、同济新村和后来的邮电新村等为城市空间特征框定了总体基调，并产生了大连路—四平路交叉口这样的较大型商业节点。该段沿线在80年代后经历了一轮较大规模的城市更新，虹临花苑外销房标志并带动了该段南部沿线的大规模房地产开发。四平大楼戏剧性的建而复拆令人印象深刻。四平路交通拥堵问题也在该段发生并日趋严重，虽然地铁和世博期间实施的道路下穿工程使之有所缓解，但总体而言，由于城市形态及其公共空间系统存在较大局限，沿线城市空间的发展矛盾仍十分突出。

(3) 南段：溧阳路—吴淞路

南段沿线是近代上海城市化的产物。石库门里弄密集，两侧的城市空间曾经是典型的高密度里弄区域及其附随的小型沿街商业设施，混杂有一些花园洋房和小型工厂。该段道路的城市更新正在进入新阶段，随着道路设施的改善，石库门逐渐被拆除，但是以瑞康里为代表的城市遗产空间亦提出了城市再开发模式的转型问题，南端西侧的虹口港地区尤为引人瞩目。该地段的交通堵塞问题同样十分突出，尤其是在四平路与海宁路（上海内城区另一条主要结构性干道）交叉的节点。交通、功能、公共空间的相互制约为这个地区的进一步发展提出巨大挑战。

■ Urban development alongside Siping Road and its spatial characteristics

Today Siping Road is an Arterial Road of more than 6 kilometers which crosses Yangpu District and Hongkou District. During the urbanization period, varied spatial typologies and qualities have been generated alongside Siping Road. We can differentiate them it into three segments of the road:

(1) Northern part: Wujiaochang to the Second North Zhongshan Road

This area had been occupied by the big "Danweis" such as the Air Force Academy of political Science and Military Departments of Air Force, as well as agriculture areas of Siping Village. This area was confronted to huge urbanization process after 1980s. In 1983, Siping Road was exlarged to a wide traffic road with landscape. It was followed by development of residential areas and commercial areas in Wujiaochang. The elevated high speed traffic road Second North Zhongshan Road was built in 1994, which further changed the original urban landscape.

(2) Middle part: the Second North Zhongshan Road - Liyang Road

Urbanization process alongside Siping Road in this part is almost completed before 1980. East-Shanghai Workers' Village, Tongji New Village and the Post and Grid New Village, which started development since 1950s, are the three main residential areas that dominated the areas alongside Siping Road and founded the spatial basis. The Dalian Road-Siping Road node as a big-scaled commercial place was given birth also during this period. The second phase of urbanization took place after 1980s and characterized by projects like Honglin Garden (residential area) which triggered the development of southern area of this part of Siping Road. The situation of traffic jam is particularly severe this part of Siping Road, although somehow released by the development of metro lines and 3-dimentional crossing road projects during EXPO Period (2004-2010). In general, conflicts exist between the traffic and the interwoven elements of urban form and spaces, which hinders the area from further upgradation.

■ **城市设计实践**

　　四平路显然是上海中心城市空间结构的重要组成之一，它是上海中心城北区的主要空间骨架。它的沿线城市空间经历了上海城市发展的所有主要阶段，因此产生了丰富的空间类型。然而，伴随着交通问题的日益突出，以及提高城市空间质量的紧迫需求，在一个如此复杂而多样的建成环境面前，如何实现系统性、针对性的沿线城市空间优化发展？城市设计者正面临着一系列的专业挑战。我们一直认为，步行环境的优化是机动车干道从发展的"边缘"回归"中心"的重要手段，也是当代城市空间品质提升的核心问题。四平路这样一条道路的步行化提升，不仅关乎道路两侧空间品质的提高，更关乎两侧城区的进一步发展。

　　城市设计的工作方式在于通过对城市要素的整合提出对既定目标的整合方案。在面对四平路具体项目的时候，以提升两侧步行品质并进一步整合优化公共空间环境为目标，我们必须回答下面几个主要问题：

　　（1）四平路沿线的步行空间的品质如何？影响品质的主要因素是什么？

　　（2）提升步行空间品质的主要结构性措施是什么？其中，"如何消除机动车道路对城市步行空间的负面影响"必须被作为最重要的问题加以解决。

　　（3）空间品质的提升如何进一步支撑两侧城区的发展以实现系统性的整体提升？通过什么手段达成目标？

　　对上述问题的回答构成了本章节后半部对各个节点设计的主要思路。设计手段普遍从"点""线""面"（干道交汇城市节点、干道沿线步行空间和干道两侧步行网络连接）三个方面展开，展现研究理论对实践的支撑。

(3) Southern part: Liyang Road-Wusong Road- the Bund

Majority of urban areas alongside the southern part of Siping Road took their form in early urbanization period of Shanghai between end of 19th Century to beginning of 20th Century, characterized by high-density Linong Housing (a typical traditional Shanghai housing form) areas, small street-side commercial buildings and a small amount of private family houses and small factory buildings. With the new urbanization take place after 1990s, the road is further enlarged with demolition of some of the surrounding Linong areas. New challenges of regeneration emerge with the difficulty to find a way of development and urban conservation. Such challenges is thus given to the regeneration projects like Rui-Kang-Li and Hongkou Harbor. The traffic jam situation is still severe in this part of Siping Road, especially at the joint to other arterial roads (like Haining Road, which is another important arterial road for Shanghai's inner city).

■ Urban Design Practices

In general, the aforementioned introduction of Siping Road and its adjacent urban areas raises a challenge for the urban design, which is that how urban design can integrate spatial elements to build a model for future in which arterial gains its importance as "center" of development, with the condition that urban space quality alongside the road is significantly enhanced. And such model also contains to ambition to expand the influence of arterial road to other areas not necessarily alongside it, to form an urban space net that covers the city entirely.

Facing Siping Road as a practical topic, the following questions shall be asked:

(1) How is the current quality of pedestrian space alongside Siping Road? What are the determining factors?

(2) What can be the structural measure to enhance quality? With answering this question, elimination of negative influence of traffic to the pedestrian space shall be treated as most important target.

(3) How the enhancement of pedestrian space quality can further influence the urban area surrounded? What design measure can be utilized?

The answering of aforementioned questions, becomes the sources of inspirations for the urban design for different parts of Siping Road in this book. These inspirations can be grouped in solutions on three aspects of "urban nodes", "sidewalks of arterial" and "street net", which are summarized and demonstrated in former chapter of this book. Theory and practices thus come together.

相关城市设计提案
Urban Design Project

五角场
WUJIAOCHANG
马宁/莱布　Benjamin Heinrich / Stefan Raab

四平路—国权路
SIPING ROAD-GUOQUAN ROAD
胡鸿源/王朝　Hu Hong Yuan / Johannes Burtscher

四平路—中山路
SIPING ROAD-ZHONGSHAN ROAD
周娜/黄迪奇　Zhou Na / Huang Diqi

同济校门
MAIN GATE OF TONGJI CAMPUS
鞠颖/王良　Ju Ying / Wang Liang

四平路—赤峰路
SIPING ROAD-CHIFENG ROAD
艾尔莎/莫非凡/爱沙/特蕾莎/迈雅/孟详皓/崔潇/谢一轩
Elsa Favier/Seyed-Movaghar/Elsa Larcher/Teresa Mayerhofer/MaijaParviaiben/Meng/Xianghao/Cui Xiao/Xie Y

四平路—大连路
SIPING ROAD-DALIAN ROAD
赵玉玲/夏琴/雷少英　Zhao Yuling / Xia Qin / Lei Shaoying

四平路—曲阳路
SIPING ROAD-QUYANG ROAD
彭何冬/迪安娜　Hedong Peng / Anna Diallo

四平路—临平北路
SIPING ROAD-LINPING NORTH ROAD
李恒晔/丽斯　Li Hengye / Clarisse Carlier

四平路—海伦路
SIPING ROAD-HAILUN ROAD
蔡兴杰/霍雅/杨丹　Cai Xingjie / Christoph Holzinger / Yang Dan

吴淞路—海宁路
WUSONG ROAD-HAINING ROAD
朱枫/阿诺　Zhu Feng / Arnaud Despretz

吴淞路—外滩
WUSONG ROAD-THE BUND
李京/穆丽莎/陈彦彤/刘泓汐　Li Jing / Lisa Mueller / Chen Yantong / Liu Hongxi

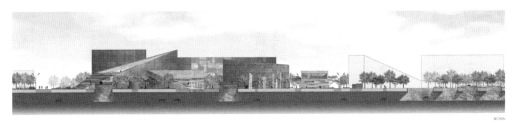

相关城市设计提案

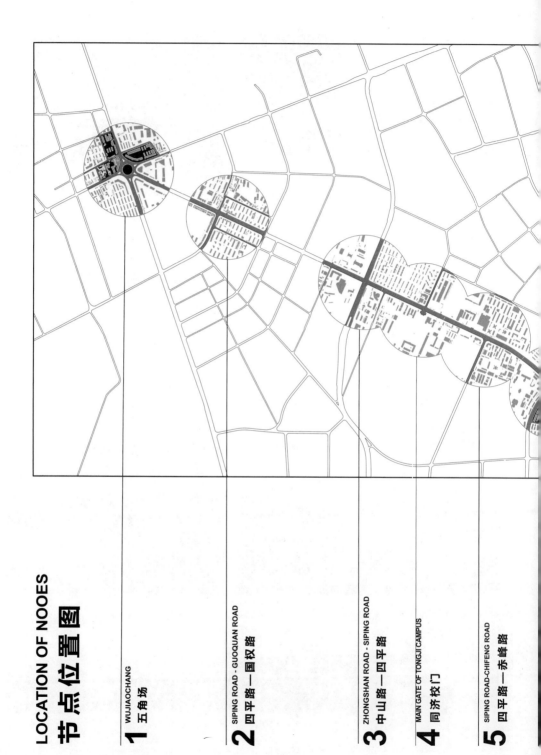

LOCATION OF NODES
节点位置图

1 WUJIAOCHANG
  五角场

2 SIPING ROAD - GUOQUAN ROAD
  四平路—国权路

3 ZHONGSHAN ROAD - SIPING ROAD
  中山路—四平路

4 MAIN GATE OF TONGJI CAMPUS
  同济校门

5 SIPING ROAD-CHIFENG ROAD
  四平路—赤峰路

7 SIPING ROAD - QUYANG ROAD 四平路—曲阳路

8 SIPING ROAD-LINPINGBEI ROAD 四平路—临平北路

9 SIPING ROAD-HAILUN ROAD 四平路—海伦路

10 WUSONG ROAD-HAINING ROAD 吴淞路—海宁路

11 WUSONG ROAD-THE BUND 吴淞路—外滩

N

PEDESTRIAN PARADISE ON ARTERIAL ROAD

WUJIAOCHANG
五角场

THE RING
"环"

Benjamin Heinrich / Stefan Raab   马宁/莱布

Planned and acting as urban center in northern region of Shanghai, Wujiaochang is situated directly in the joining point of five arterial roads (Siping Road, Handan Road, Songhu Road, Xiangyin Road and Huangxing Road). The public space and pedestrian system is thus cut by the five arterials into very fragmentary state. The environment quality of this area is low, which hinders the area for further development. The current pedestrian linkage between the building blocks at the joining point can only be realized through underground passages which are connected with metro station.

Our design brings forward a vision that the whole area surrounding Wujiaochang must be turned into a slow-traffic area with very high pedestrian quality, which can only be realized through an integrative re-planning of traffic system. Taking the reference of Ringstrasse in Vienna, the outskirt roads of Wujiaochang (existing ones)are interconnected to form a single-direction traffic ring which prevents the passing traffic from crossing the center of Wujiaochang and makes traffic pressure within the area much lowered. The width of five arterial roads can be thus reduced and make place for pedestrian and mix-use development.

　　五角场是上海北区的城市副中心。它以在交通方面的重要性出名——五条主干道在此交汇（四平路、邯郸路、淞沪路、翔殷路和黄兴路）。这个地区的步行系统被五条干道切割，地面步行环境碎片化严重，恶劣环境难以支撑它作为城市副中心的功能需求。目前的主要步行连接被置于地下层，与地铁站厅连接，由商场和人行道上的扶梯上地面层。

　　我们的设计尝试通过地面交通系统的整体改造解决以上问题。借鉴维也纳一期和环城大道的做法，将主要交通设置于外围的单向交通环道，避免过境车流进入核心区，从而把以五角场为中心的接近一个平方公里的区域转换成慢行和步行混合的区域，形成公共空间网络。由于区域内部交通压力减小，原来的主干道路宽度减小，有空间对两侧城区混合功能模式进行开发设计。

BIRDVIEW ON WUJIAOCHANG NODE
五角场城市节点鸟瞰

Research Scope of Wujiaochang Node
五角场节点研究区域

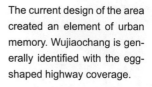

The current design of the area created an element of urban memory. Wujiaochang is generally identified with the egg-shaped highway coverage.

当前该区域的设计制造出了城市记忆的某种元素，提及五角场，便会联系到包裹其高架道路的鸡蛋状覆盖物。

relationship to the "sub-centers" of shanghai
与上海"亚中心"的关系

The major roads crossing the area are dividing it in five parts and segregate the functions of the buildings from each other. No direct connection is existent.

穿越该区域的城市主干道在此分流为五部分，并将此处建筑群的功能互相分离，联系甚少。

segregation due to major traffic roads
主要交通干道的隔离

The pedestrian|human itself is abandoned under the earth with to an underground plaza in the center and a shopping-mall leading to the north.

该区域内的步行者被"抛弃"于与中心地下广场相连的各个商业购物中心和地下步行街。

problem focus in the collision point
碰撞点中的问题焦点

PROJECT ONE: THE RING　　　　　　　　　　　　CHAPTER　8

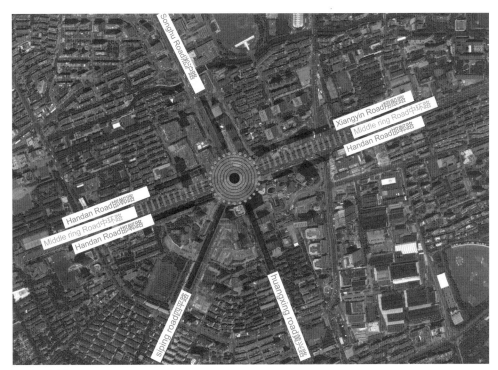

Five Arterial Streets
汇聚的五条干道

Siping Road　四平路

Huangxing Road　黄兴路

Songhu Road　淞沪路

Handan Road　邯郸路

作品一："环"　　　　　95

PEDESTRIAN PARADISE ON ARTERIAL ROAD

CASE:The center of Vienna is surrounded by a ring-road, former location of the city wall, which eases the traffic and redirect it, to provide a traffic pacified zone.
案例：维也纳中心由指环路环绕，缓解过境压力。

The concept of the ring-road shall be translated to the building-site of Wujiaochang. Most of the intended "ring-road"-roads are already wide enough to bear the expected traffic.
案例：在五角场节点中运用"环道"的概念。大多数已有"环路"已经足够宽，能承担预期的交通压力。

PROJECT ONE: THE RING CHAPTER 8

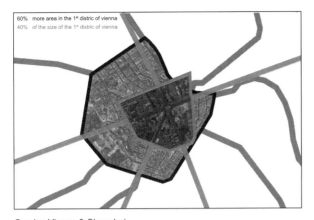

Overlay Vienna & Shanghai
覆盖维也纳和上海

The overlay of the two systems highlights the similarities of the cases and illustrates the difference in size (amount of traffic is quite similar).

将上海与维也纳两个系统相互叠加，反映出了两个案例的类似性以及在尺度上的差异性（两者交通流量相对接近）。

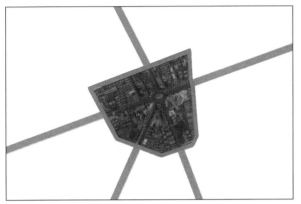

Cutout of Wujiaochang | circleroad inner area
五角场环路区域内切口

By taking the traffic collision out of the center, the major problem will be divided in smaller sub-problems.

通过将交通冲突点移出五角场中心区域，将交通问题分散化。

Problem Distribution | one major issue separated in five smaller problems
一个问题分解成五个小问题

The sub-problems are now located at the intersection of the ring road with the five main roads. (going into the direction of Wujiaochang).

程度相对较轻的交通问题被转化至五个主干道与区域环路交叉口处消解。

作品一："环"   97

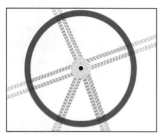

Scope of Research
研究范围

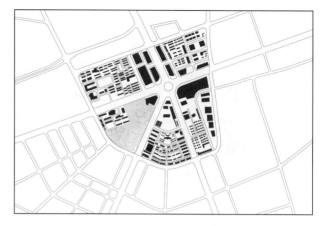

Existing Road System
现状交通系统

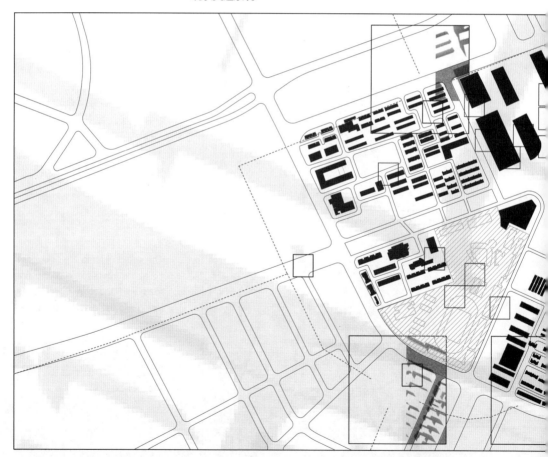

Redesigned Inner Road System
重新设计的内部道路系统

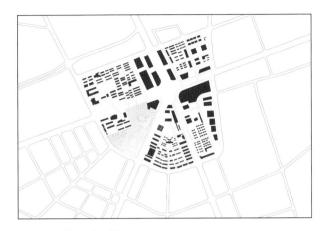

Redesigned Inner Road System
重新设计的内部道路系统

The new system will be a pacified central area with a speed limit of 30 km/h. The parking lots inside will be reserved for the residents. For visitors new underground parking spaces will be built at the boarders of the area.

新系统将成为一个限速30公里/时的慢行中心区域。内部停车场将只对周边居民开放，并将在该区域的边界处为观光者建设新的地下空间。

Due to the ease of the traffic inside of the ring-road, the current streets become unnecessary and will be reconfigured|reevaluated according to the expected amount of traffic.

鉴于环路内部交通量的减少，当前的街道尺寸会显得过大，届时需要根据实际交通量进行评估与改造。

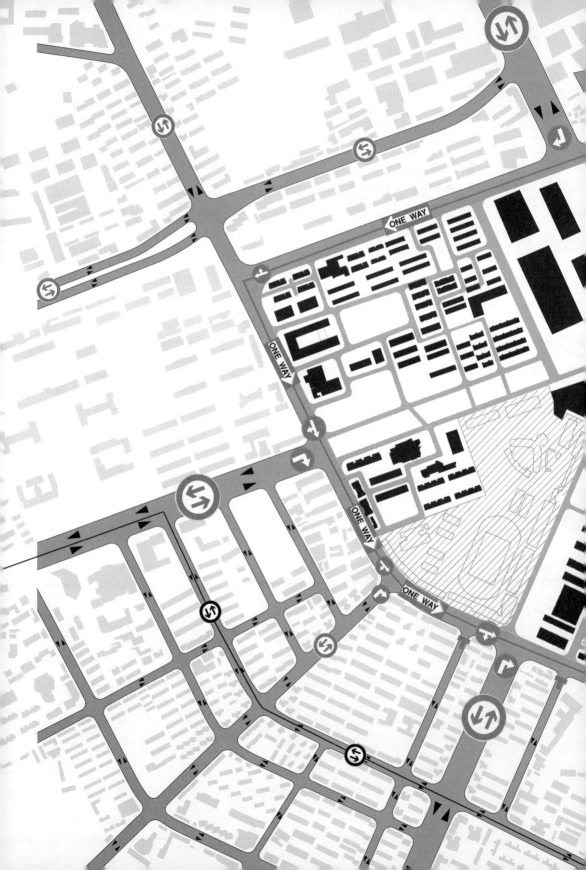

PEDESTRIAN PARADISE ON ARTERIAL ROAD

Concept Drawing
概念图

Guoding Road
国定路

Section | Before and After
剖面图 | 改造前后

View in direction of Wujiaochang
从五角场方向看

102　　　步行与干道的合集

PROJECT ONE: THE RING CHAPTER 8

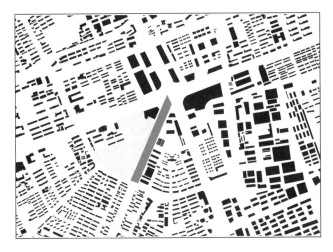

Concept Drawing
概念图

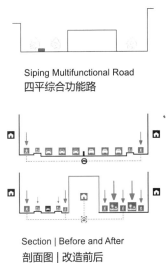

Siping Multifunctional Road
四平综合功能路

Section | Before and After
剖面图 | 改造前后

View in direction of Wujiaochang
从五角场方向看

作品一："环" 103

PEDESTRIAN PARADISE ON ARTERIAL ROAD

Concept Drawing
概念图

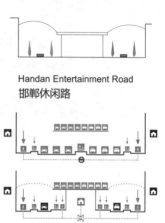

Handan Entertainment Road
邯郸休闲路

Section | Before and After
剖面图 | 改造前后

View in direction of Wujiaochang
从五角场方向看

PROJECT ONE: THE RING   CHAPTER 8

Concept Drawing
概念图

Songhu Park Road
淞沪公园路

Section | Before and After
剖面图 | 改造前后

View in direction of Wujiaochang
从五角场方向看

作品一："环"   105

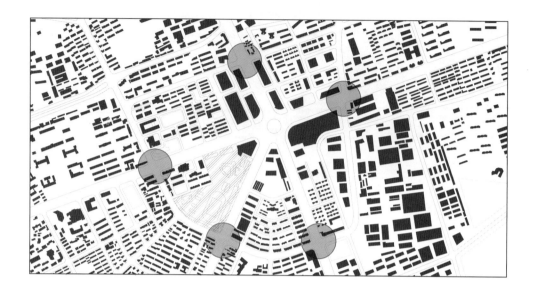

By excluding the traffic from the inner ring area, sub-problems have been created at the intersection points of the ring-road with the major roads. These new zones need to be solved in terms of two major issues they are facing, on the one hand the pedestrian system and on the other hand the guidance of the traffic to avoid confusion about the direction of using the one way circle road.

通过将机动交通移出环岛区域，子问题已经被置于主要道路的环形路交叉点了。这些新的区域需要解决两个重要问题：一是步行系统，二是对机动交通的引导，以避免因使用单行道而产生的冲突。

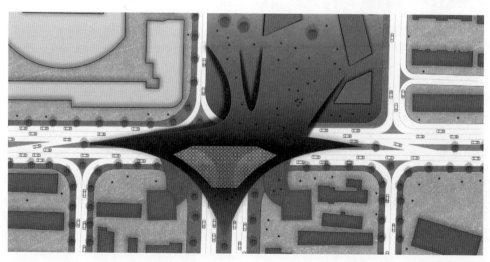

PROJECT ONE: THE RING   CHAPTER 8

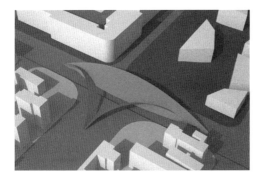

The first step is the guidance of the traffic via a road separation system, which should take the flow of the traffic and guide it directly in the intended direction.

通过道路交通分离系统的引导。根据交通流量实时引导它直接进入预定的方向。

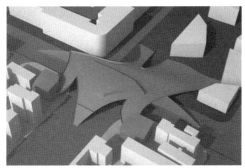

The conceptual "grid" is superimposed with the design concept. The overall design looks [material-wise] naturally grown on the up and artificial on the downside.

概念"网格"叠加的设计概念。整体设计表面貌似自然增长，实则遵循其内在秩序。

By approaching the structure on street level, guidance lights help the car drivers to understand their options of driving, which are: on the ring-road, the choice of following the ring or exiting at any major road and from the major roads, the only possibility is to enter the ring counter-clockwise.

通过符合街道尺度的构造，引导灯帮助汽车司机了解他们应选择的驾驶方向。本节主要展示其构造示意，展示主要内部空间是如何在视觉上强调桥上与桥下的连接向导的。

PEDESTRIAN PARADISE ON ARTERIAL ROAD

步行与干道的合集

作品一:"环"

PEDESTRIAN PARADISE ON ARTERIAL ROAD

② SIPING ROAD - GUOQUAN ROAD
四平路—国权路

THE LINK
连接

Hu Hong Yuan/Johannes Burtscher   胡鸿源/王朝

The main goal of "link" is to connect Fudan University with the city. On the one hand Guoquan Road will be transformed into a pedestrian road which will provide a safe passage for all the students from the metro to the university. There will be several functions related to the university. Specialized shops, cafe shops , bars and restaurants will vivid the street and transform the whole area into a public space as well for students as for local people . On the other hand the metro station will connect the divide Guoquan Road and create an easy passage to cross Siping Road. The Courtyards on both sides of Siping Road will change the existing metro station into a nice place not only to pass but also to stay. The yard will be characterised by greenery, street furnitures and shops. A land mark will be the public library which gives the main courtyard unique character.

"连接"的主要目标是将复旦大学与城市联系起来。 一方面，国权路被转换成一个人行道路，为所有的学生提供一条从地铁站到大学的安全通道。这里设置了一些与大学生活相关的功能设施，商店、咖啡店、酒吧和餐馆等能使整个街变得生动有活力,而且对学生及生活在周边的居民来说这个地区将变成一个公共空间；另一方面，地铁站将分裂的国权路连接起来，创建一条更方便通行的四平路。四平路两边的庭院将现有的地铁站变成一个非常漂亮的地方，在这里，人们不仅可以穿行，也可以停留下来休息。院子以绿色植物、街道设施和商店为主。地标性公共图书馆使主要的院子有独特的标识。

BIRDVIEW ON SIPING ROAD-GUOQUAN ROAD NODE
四平路—国权路城市节点鸟瞰

Continue The Structure
保留结构

■ New　新的
▥ Retrofits/Integration　更新/集成

Function
功能

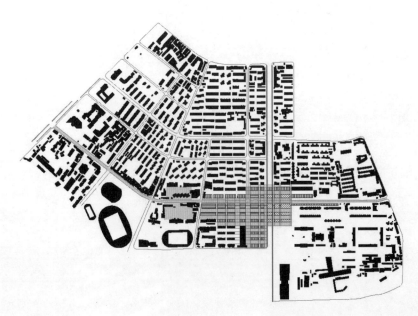

▥ Residential/Residential Mixed Use　居住区/居住功能混合
▥ Office/Office Mixed Use　办公/办公功能混合

Creating Space
创造空间

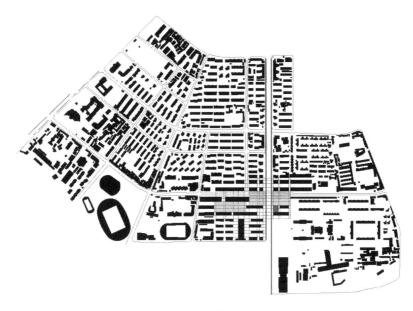

▦ Plaza 广场
■ Landmark 地标

Concept
概念

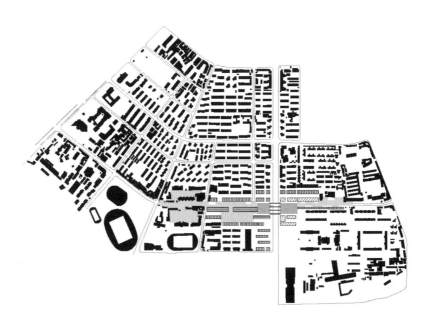

▦ Residential/Residential Mixed Use 居住区/居住功能混合
▨ Office/Office Mixed Use 办公/办公功能混合

PEDESTRIAN PARADISE ON ARTERIAL ROAD

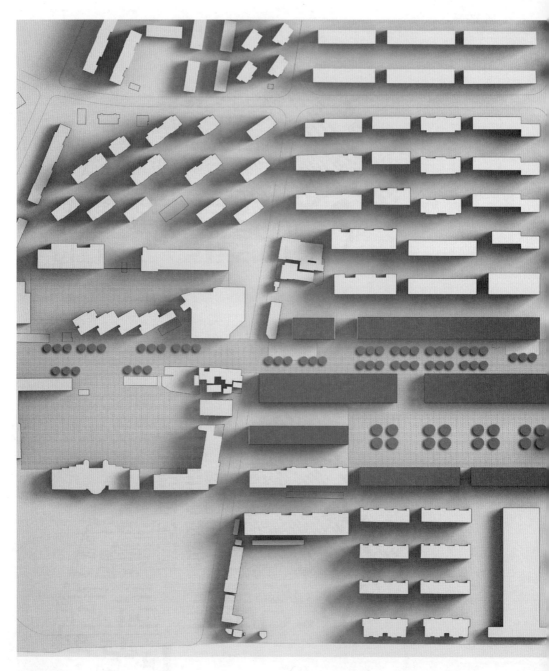

Master Plan
总平面

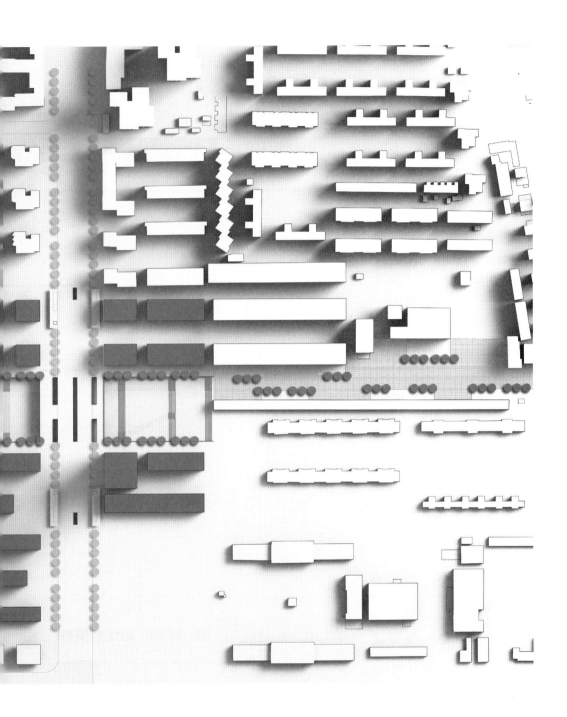

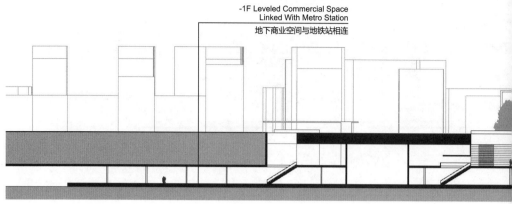

Section A
剖面 A

Plan of Underground Passage
地下通道平面图

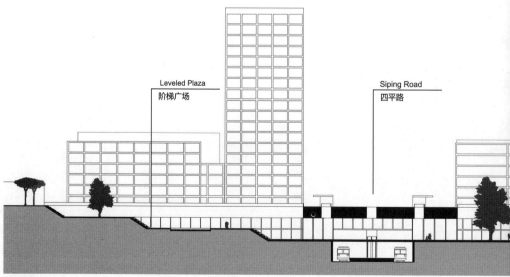

Section B
剖面 B

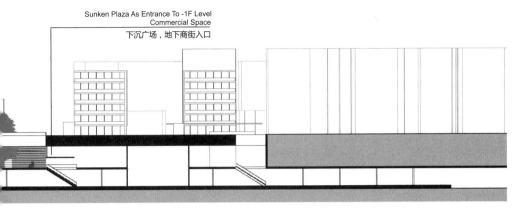

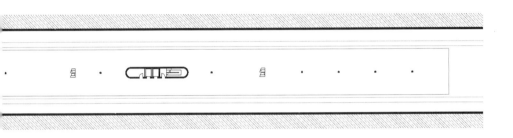

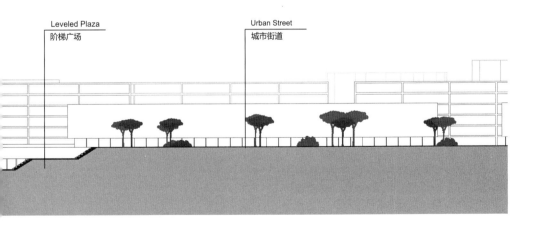

PEDESTRIAN PARADISE ON ARTERIAL ROAD

③ SIPING ROAD- ZHONGSHAN ROAD
四平路—中山路

RIVERSIDE PARK
滨水公园

Zhou Na / Huang Diqi   周娜/黄迪奇

The Second North Zhongshan Road and the elevated express road (north part of inner ring road of Shanghai), crosses here with Siping Road and forms very complicated node here. All these roads are very wide with heavy traffic which makes it very hard for passengers to cross. All building blocks alongside the node build walls to face the road, which makes the environment even unfriendly to people. The river alongside the Second North Zhongshan Road in its north is even not possible to discover from passengers perspectives.

Our design develops a system in multiple levels to link the separated blocks into one entity, which cannot be affected by motor traffic. All connections lead to the riverside park, which is designed with platform directly situated on the water level as well as many facilities that can support urban activities in the park. The park acts as the core public space for the students, professors, residents and employees that live or work in surrounding area.

中山北二路及覆盖于其上的的内环路高架北段与四平路交汇，形成了一个非常复杂的节点。道路交通量大，很难穿越，南侧的同济大学及附属用地和北侧的产业园区沿路都设置围墙，进一步恶化了这个地块的步行环境质量，人们甚至难以察觉中山北二路北侧一条河流的存在。

城市设计将道路两侧的各个地块通过天桥和地下通道联系起来，步行连接可以不被机动车交通影响。沿河地面的标高被降低到亲水的平面，形成下沉的景观广场，这里将成为一个城市公园。沿水岸设置很多公共功能，附近的学生、教师、同济新村的居民、创意园区的员工可以在这里聚会和活动。

BIRDVIEW ON SIPING ROAD-ZHONGSHAN ROAD NODE
四平路—中山路城市节点鸟瞰

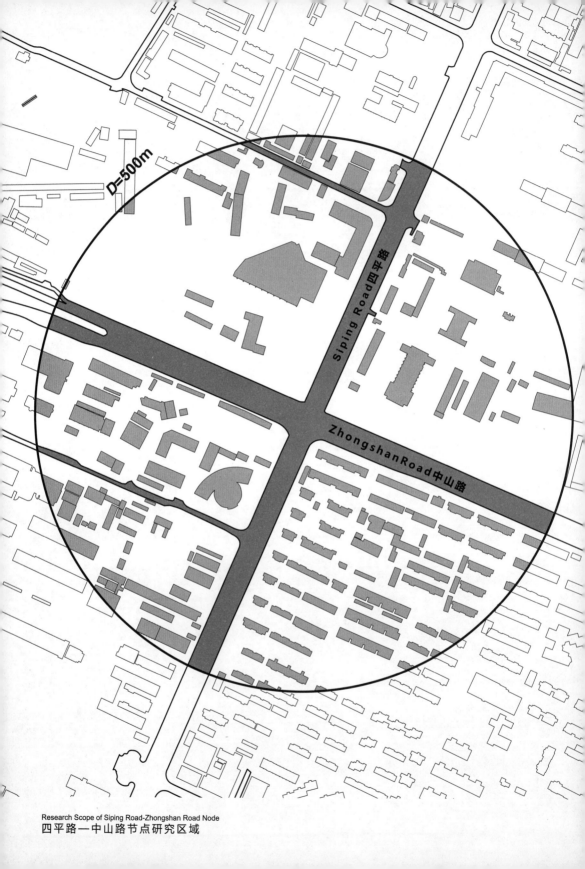

Research Scope of Siping Road-Zhongshan Road Node
四平路—中山路节点研究区域

Public Space Under Elevated Road
高架桥下的公共空间

Section A
剖面 A

Section B
剖面 B

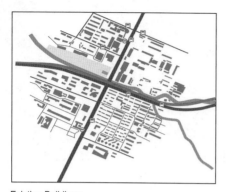

Existing Buildings
现状区位

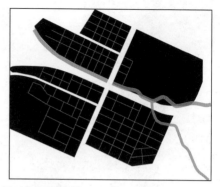

Planned Street Net
规划的街区网络

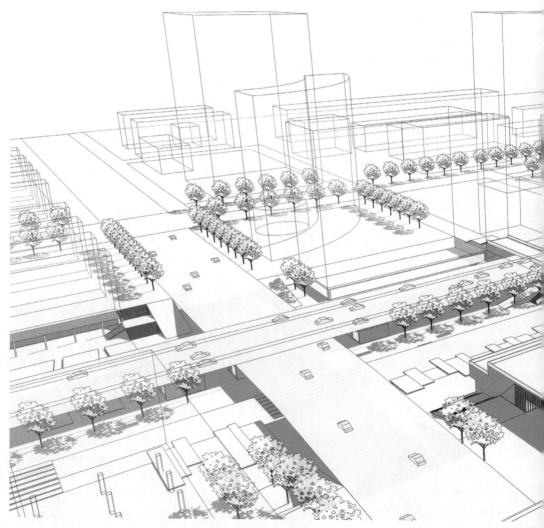

PROJECT EHREE : RIVERSIDE PARK CHAPTER 8

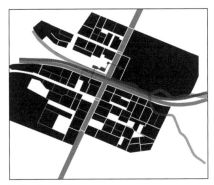

Public Space Net
公共空间节点

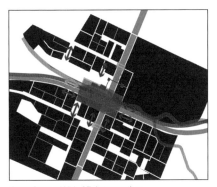

Core Space With 3D Intersection
立体交叉的核心

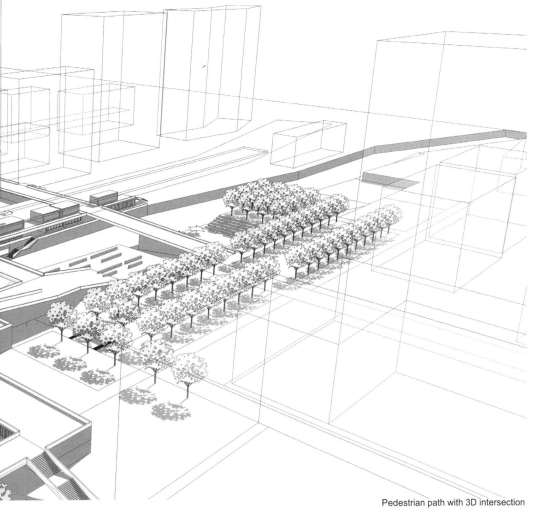

Pedestrian path with 3D intersection
立体交叉的步行路径

作品三：滨水公园

④ MAIN GATE OF TONGJI CAMPUS
同济校门

PUBLIC SPACE RING
环形公共空间

Ju Ying / Wang Liang   鞠颖/王良

Zhangwu Road, which acts as active community-level commercial street, has its joining point here with Siping Road. This is also the place where the main gate of Tongji University is situated. Everyday thousands of student enter and exit from this gate and cross Siping Road to Zhangwu Road for their leisure and living. Safety, comfort and function shall all be considered for the urban design of this node.

We design here elevated pedestrian system in form of a ring, which links the important buildings surrounding this node together (the 2 office buildings in Tongji Campus and the 2 high-rise office buildings on the other side of Siping Road) as well as the important places such as the main gate plaza of Tongji, beginning of Zhangwu Road and those exits of Metro Station. This facility is not only a tool to help people cross road, but an urban complex which integrate function such as public classrooms, small library, coffee-bars and restaurants. The "in-between space" is thus becoming a place with vitality and fun.

　　作为区域主要商业街道并且比较具有活力的彰武路与四平路交汇。这里也是同济大学的正校门所在地，每日大量学生需要从校门进出，往返于学校与住宿区之间，他们都面临着穿越四平路这条机动车交通繁忙的干道的问题。因此，道路的安全性、舒适性、功能都必须得到提高。

　　在这里我们设置了一个环形的二层步行系统，它将节点四周的主要建筑联系起来（同济大学的两栋行政办公楼，四平路对面的两座高层办公楼），同时也把校门、地铁站、彰武路口这些重要的场所联接起来。它不仅是一个帮助行人过马路的设施，其自身也是一个公共功能的综合体（学生公共教室、小型图书馆、咖啡酒吧、餐厅），让这个"之间空间"变成充满活力和趣味的场所。

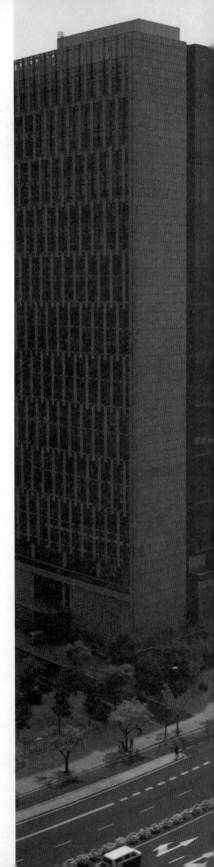

BIRDVIEW ON TONGJI GATE NODE
同济校门城市节点鸟瞰

Research Scope of Tongji Gate Node
同济校门节点研究区域

PEDESTRIAN PARADISE ON ARTERIAL ROAD

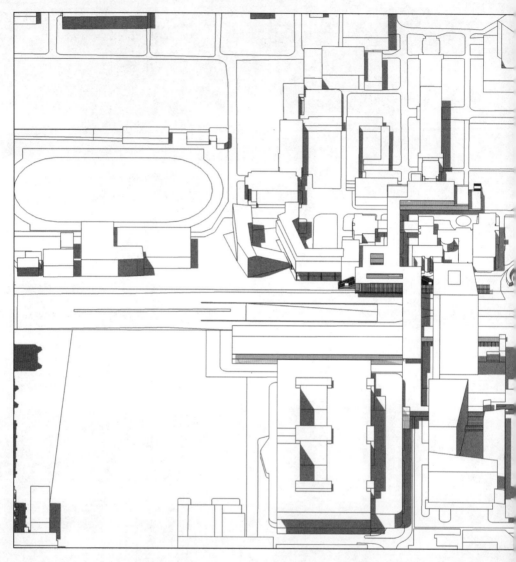

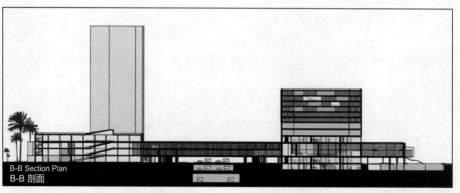

B-B Section Plan
B-B 剖面

PROJECT FOUR : PUBLIC SPACE RING　　　　　　　　　　CHAPTER 8

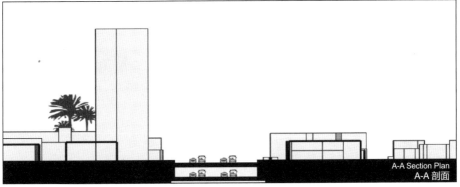

A-A Section Plan
A-A 剖面

作品四：环形公共空间

PEDESTRIAN PARADISE ON ARTERIAL ROAD

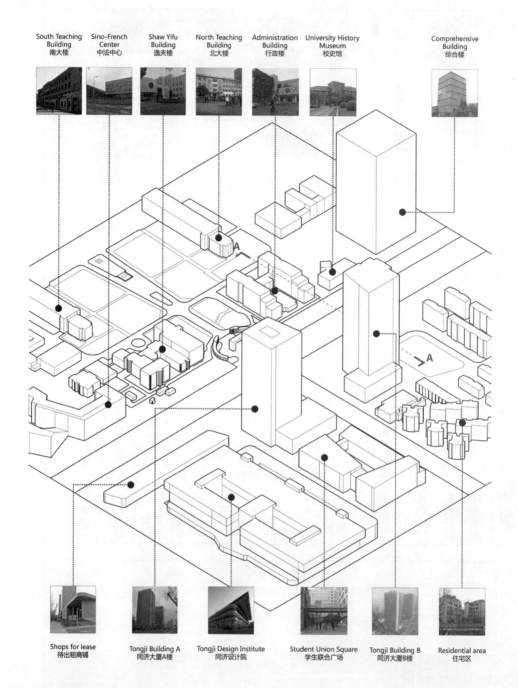

Layout of Existing Buildings
现状建筑布局

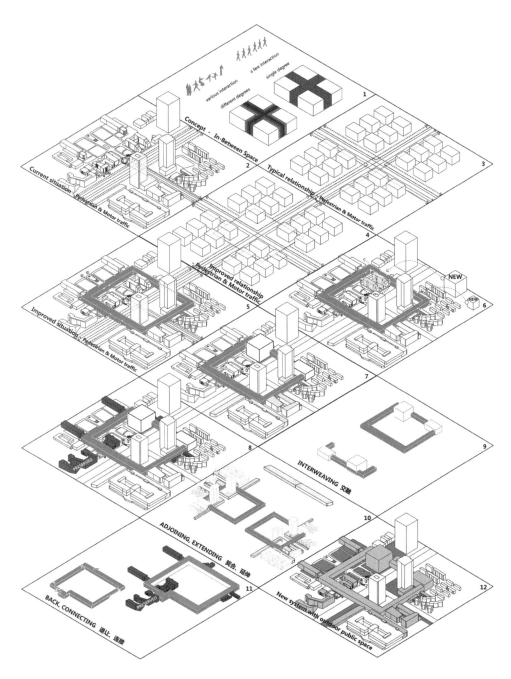

Concept Diagram
概念图解

PEDESTRIAN PARADISE ON ARTERIAL ROAD

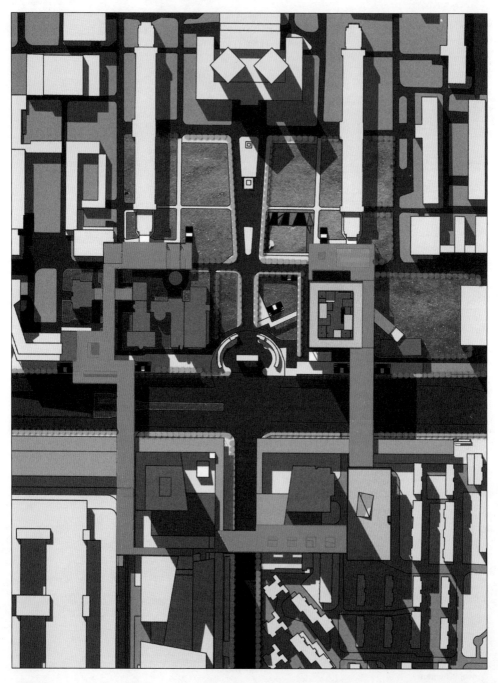

Master Plan
总平面

PROJECT FOUR : PUBLIC SPACE RING   CHAPTER 8

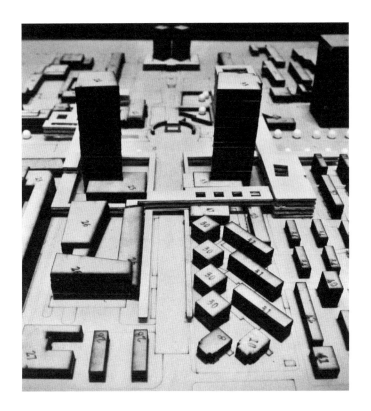

Planned Street Sidewalk Space
规划的沿街空间

作品四：环形公共空间

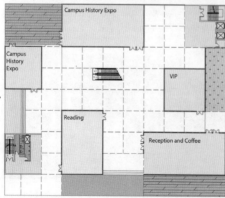

Underground Floor Plan 1:500
地下层平面 1:500

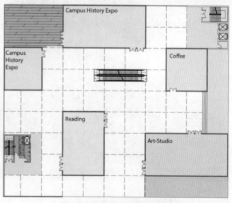

Ground Floor Plan 1:500
地面层平面 1:500

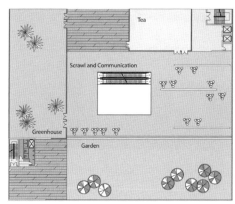

First to Second Floor Plan　1:500
一至二层平面　1:500

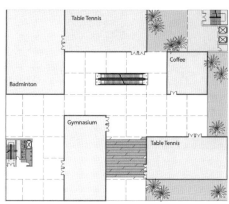

Third Floor Plan　1:500
三层平面　1:500

Birdview on Tongji Gate Node
同济校门节点鸟瞰

PEDESTRIAN PARADISE ON ARTERIAL ROAD

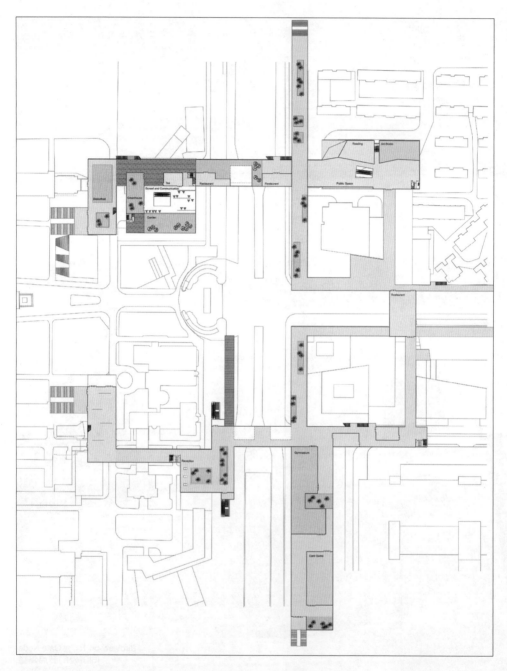

First Floor Of The System
一层平面

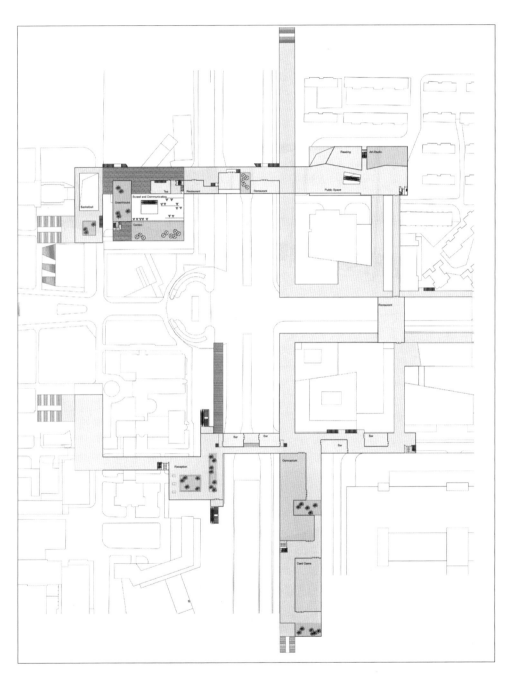

Second Floor Of The System
二层平面

PEDESTRIAN PARADISE ON ARTERIAL ROAD

⑤ SIPING ROAD - CHIFENG ROAD
四平路—赤峰路

The cross between Siping Road and Chifeng Road is about 500 meter away from that between Siping Road and Zhangwu Road, which is a very long distance for passengers. This part of road is bordered by walls from both sides (on the one side it is the wall of campus, while on the other side it is the wall of Tongji Design Institute), which cannot provide function to the sidewalks. The arterial road was enlarged in past years to give way to the entrance of the tunnel, which makes the sidewalks even narrower.

赤峰路口与彰武路口之间的距离约为500米，对于行人而言这是一个很长的距离。这个区域的四平路是一个被围墙限定的区域（南侧的设计院围墙和北侧的学校围墙），两侧的围墙都未能给予人行道功能上的支持。由于加建下穿道路被拓宽，致使两侧人行道非常狭窄，人行体验非常恶劣。

BIRDVIEW ON SIPING ROAD-CHIFENG ROAD NODE
四平路—赤峰路节点鸟瞰

Research Scope of Siping Road-Chifeng Road Node
四平路—赤峰路节点研究区域

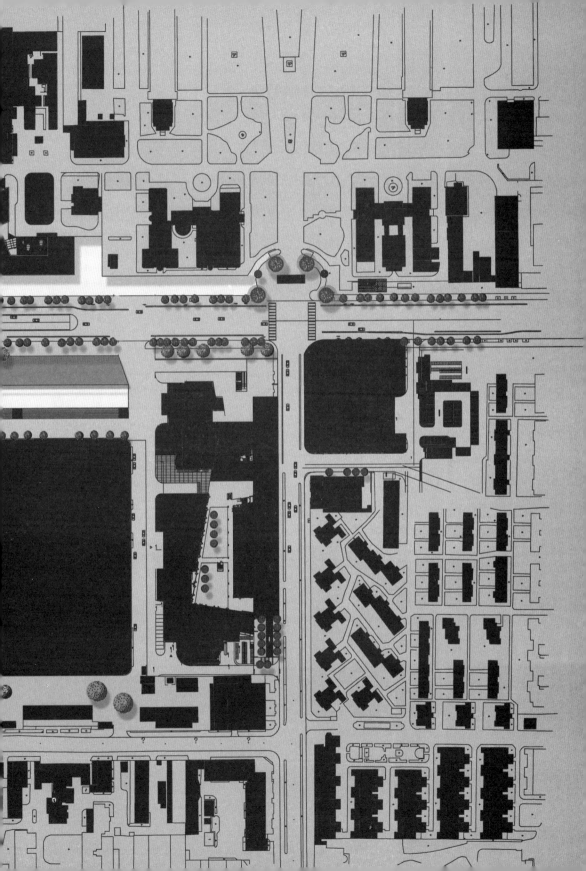

PROJECT A BURIED STRATUM
方案A 被遗忘的地下层

Elsa Favier / Seyed-Movaghar Apameh
艾尔莎/莫非凡

PROJECT FIVE A : BURIED STRATUM   CHAPTER 8

This scheme proposes to open up the building properties in both sides of Siping Road and then create huge public space composed of park and plazas, which are on -3 meter level below the ground. In spite of this, levels of motorize vehicles and pedestrians is divided. Quality of public space is thus ensured and also passengers can take the advantages of different levels to cross the arterial road.

城市设计提出开放两侧的地块（校园和设计院），并将其转变为开放式管理，将两侧地块的基面下沉3米，形成大型下沉公园。下沉的基面帮助行人可以较容易地穿越道路，同时把机动车对公共空间的影响降低。公园的外侧成为混合功能城市开发的基地。

PEDESTRIAN PARADISE ON ARTERIAL ROAD

PEDESTRIAN PARADISE ON ARTERIAL ROAD

PROJECTB CONNECT/INTERACT
方案B 连接与互动

Elsa Larcher / Teresa Mayerhofer / Maija Parviainen
爱沙/特蕾莎/迈雅

PROJECT FIVE B : CONNECT/INTERACT　　　　　　　　　CHAPTER 8

The scheme plans a huge pedestrian net covering both sides of Siping Road, treating the site on the other side of Siping Road (against Tongji Campus) as extension of campus development which is planned to be developed into places contain learning, reading, relaxation and accommodation. Pedestrian street, plaza, garden and small-scaled buildings compose intimate urban area.

这个方案里，一张步行网络被设计以覆盖四平路两侧的街区，并把四平路的另一侧地块规划成校园的扩展用地以容纳学生的学习、读书、休闲和住宿。步行街道、广场、花园（相互连接的地下、地面和屋顶花园）和小尺度的建筑形成惬意的城区。四平路作为主要服务过境交通的道路被设计为隧道。

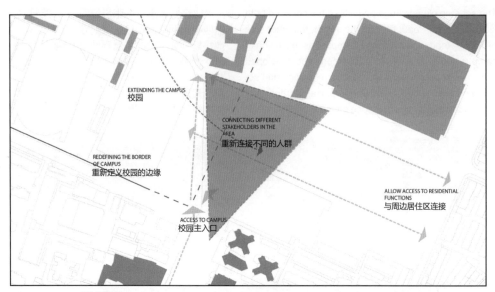

Concept Digram 概念图示

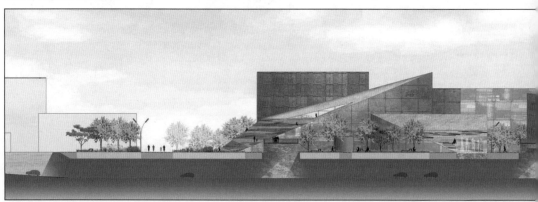

Elevation 立面

Section 剖面

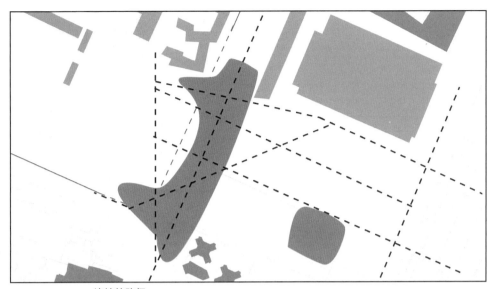
Path of Connection 连接的路径

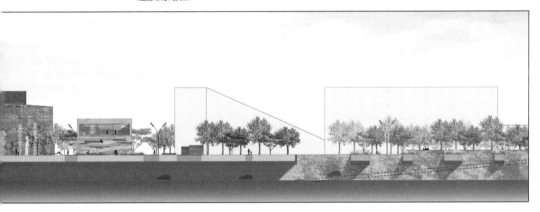

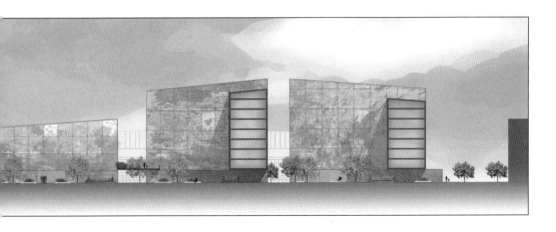

PEDESTRIAN PARADISE ON ARTERIAL ROAD

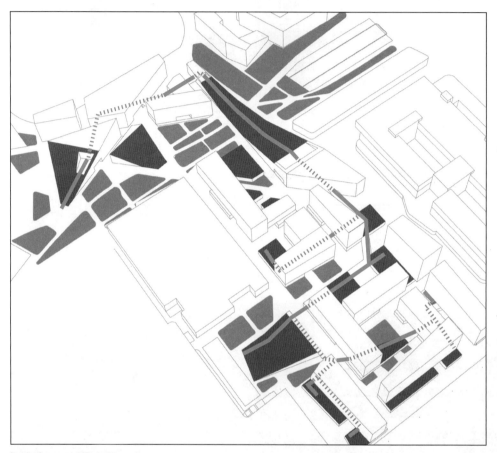

Public Space and Their Connections
公共空间及其连接

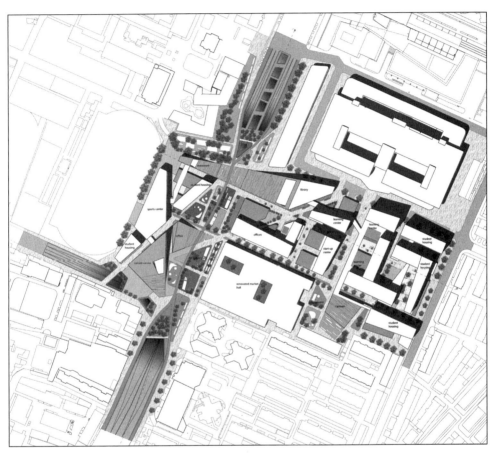

Master Plan
总平面

作品五B：连接与互动

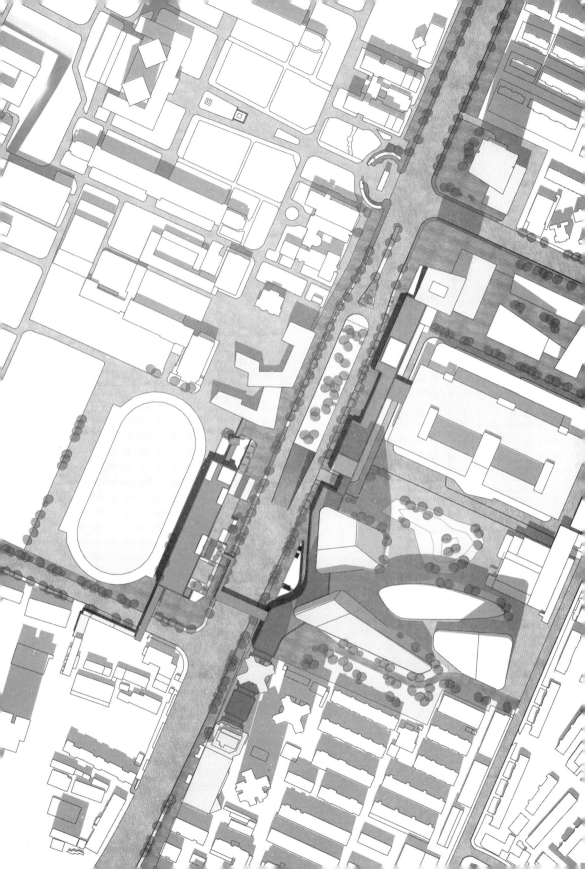

PROJECTC SPACE GENERATION
方案C 空间再生

Meng Xianghao / Cui Xiao / Xie Yixuan
孟详皓/崔潇/谢一轩

PROJECT FIVE C : SPACE GENERATION　　　　　　　　　　　CHAPTER 8

First Step :
The site area two sidewalks of Siping Road
第一步：
四平路站点区域人行步道

Second Step :
Functions along with the two path
第二步：
沿路径设置功能

Third Step :
Bottom overhead and roof garden open to sidewalk
第三步：
底层架空，屋顶花园向沿街人行道开放

Fourth Step :
Make connections between Siping Road and Chifeng Road
第四步：
四平路和赤峰路之间建立联系

Fifth Step :
Square infront Design No.1 and big slop
第五步：
在1号设计区域和大斜坡之间设置广场

Sixth Step :
Public space make the four office buildings in series and roof garden open to the pitch
第六步：
公共空间把这四幢建筑连为一体，屋顶花园向球场开放

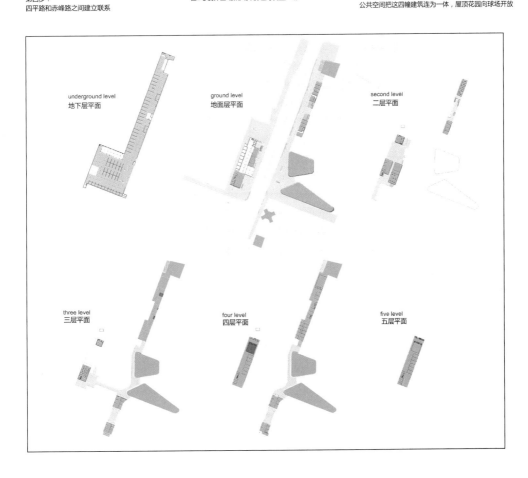

作品五C：空间再生　　　　　　　　　　　167

PEDESTRIAN PARADISE ON ARTERIAL ROAD

This scheme attempts to establish effective street building surfaces to improve the spatial quality and function of the sidewalks. In the southern side of Siping Road, it is realized by renovation and extension of existing 2-storied building with refilled urban functions, while in the southern part the wall of campus is partially opened the rest redefined by public buildings to be added.

这个方案强调了这段道路上城市界面的塑造，以之取代目前的围墙。沿设计院一侧的两层建筑被改造向两侧延伸，设置各种公共功能，形成有效的街道界面，令人行道空间的舒适度和活力得以提升。沿校园一侧的界面部分完全开放，由新建的公共建筑来加以界定。

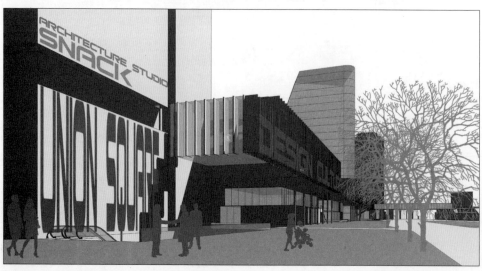

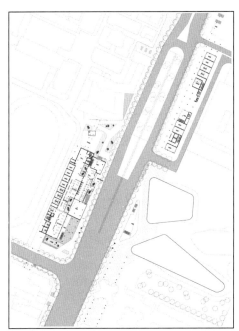

First Floor
一层平面

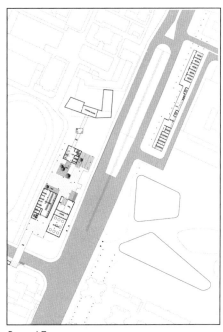

Second Foor
二层平面

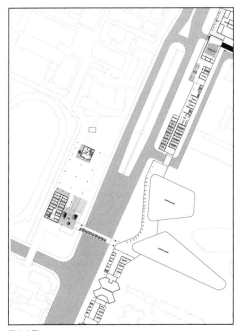

Third Floor
三层平面

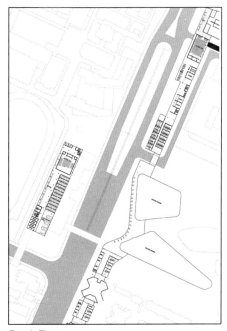

Fourth Floor
四层平面

PEDESTRIAN PARADISE ON ARTERIAL ROAD

⑥ SIPING ROAD - DALIAN ROAD
四平路—大连路

SKY PARK
空中公园

Zhou Yuling / Xia Qin / Lei Shaoying   赵玉玲/夏琴/雷少英

Dalian road is one of the important arterial roads in northern Shanghai to link the direction of east and west. The joining point of Dalian Road and Siping Road is very important in traffic structure of Shanghai and it thus triggered agglomeration of high density urban development in its surrounding area, with hotels, cinemas and transit station for two metro lines. However, the two arterial road makes the inconvenience of pedestrian linkage between the blocks. The quality of public space cannot support the further upgradation of the area.

The urban design put forwards the concept of "sky park", which establishes big platform directly above the joining point of arterial roads, which connects the existing high-rise buildings in multiple level. This platform is a park and urban space for all surrounding areas. People can enter the park with escalators and elevators linked to the ground level or metro station, while also through the internal elevators of the buildings surrounded. The platform itself is a multi-storied building which contains public functions that the further development of this area demands.

大连路是上海北区联系城市东西向交通的主要干道，它与四平路的交叉口是城市重要的交通节点，并引发早期的功能集聚，如酒店、影院等，并有两条地铁线在此交汇。道路的交汇带来建筑地块之间联系不便，步行环境和公共空间难以形成，阻碍区域进一步发展。

城市设计提出了"空中公园"的愿景。通过直接在交叉口上设置巨构平台形成不受机动车干扰的公共空间，该平台将原有的建筑进行多层次的连接，并通过扶梯和电梯将地面与地铁站的人群导入。这个平台本身是一个多层建筑，容纳了社区进一步发展需要的各种功能空间。

BIRDVIEW ON SIPING ROAD-DALIAN ROAD NODE
四平路—大连路节点鸟瞰

Research Scope of Siping Road-Dalian Road Node
四平路—大连路节点研究区域

View from Sky Park
空中公园上的场景

It's happening in all mega-cities like Shanghai: People, space and physical existance are pressed into an intensive high-density city area. They collide and they connect, breaking up the regular connection mode on one level, building up a multi-leveled connection system in a 3-dimnensional way.

它发生在所有像上海这样的大城市中：人、空间和物体的存在被压缩成一个密集的高密度城市地区。他们的碰撞与连接打破了在一个层面上的常规的连接方式，在三维空间建立一个多层次的连接系统。

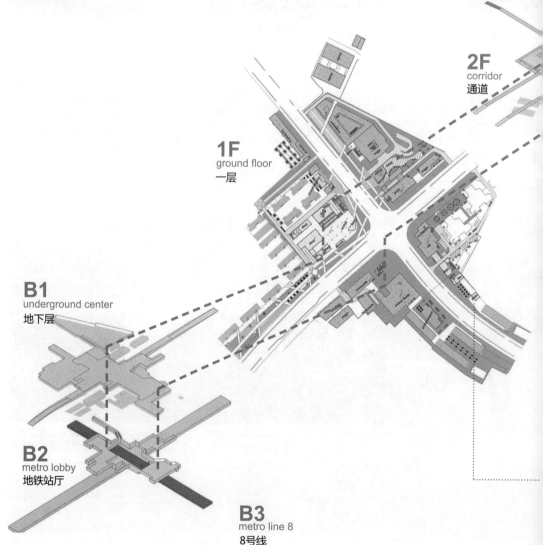

PROJECT SIX : SKY PARK　　　　　　　　CHAPTER 8

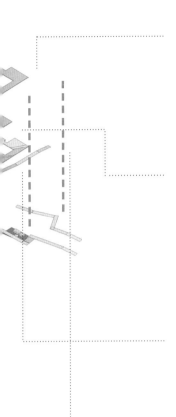

Analysis of Circulation
流线分析图

作品六：空中公园　　　　　　　　179

PEDESTRIAN PARADISE ON ARTERIAL ROAD

作品六：空中公园

PEDESTRIAN PARADISE ON ARTERIAL ROAD

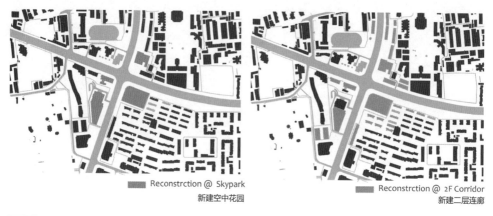

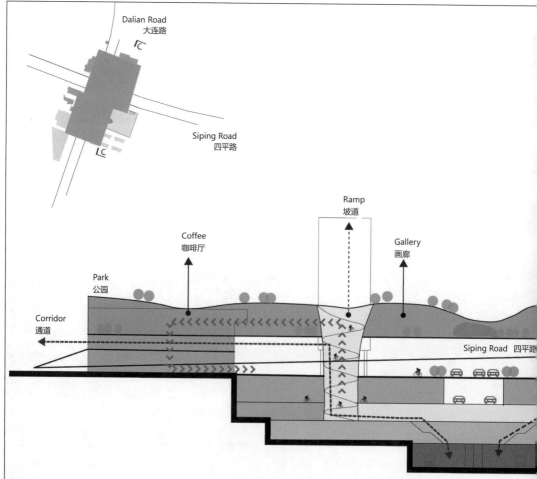

SECTION
剖面图

步行与干道的合集

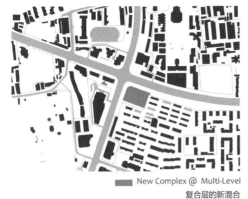

New Complex @ Multi-Level
复合层的新混合

Skypark @ Multi-Level
复合层空中花园

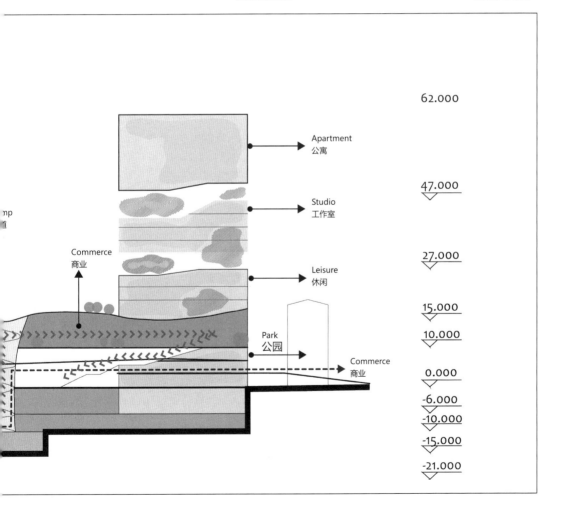

PEDESTRIAN PARADISE ON ARTERIAL ROAD

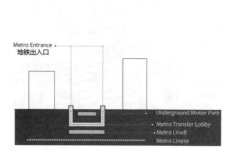
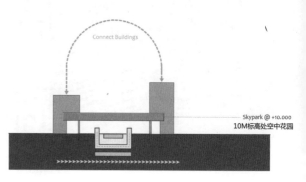

Section
剖面图

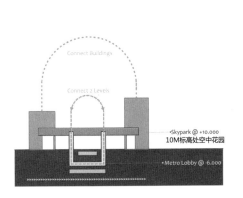
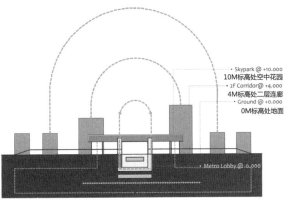

作品六：空中公园

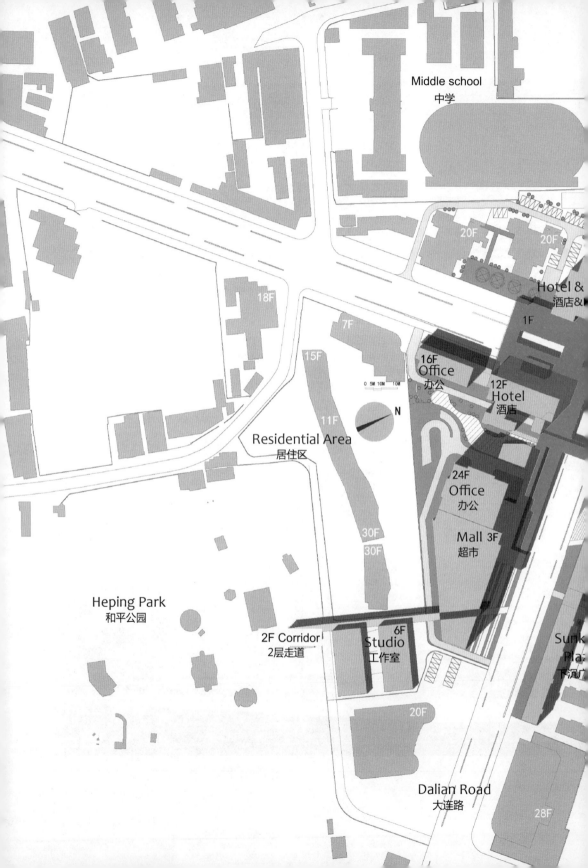

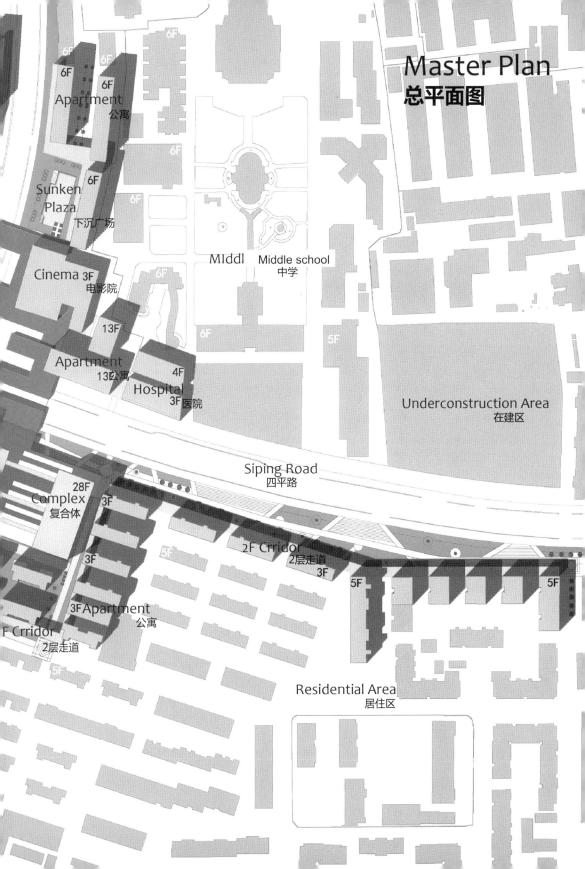

PEDESTRIAN PARADISE ON ARTERIAL ROAD

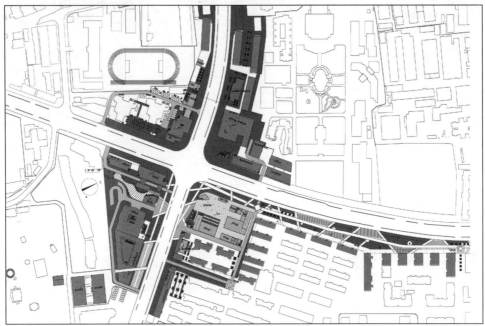

First Floor
一层平面

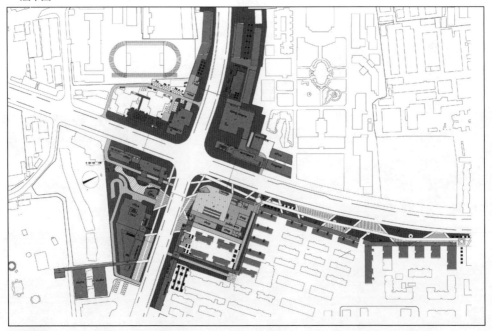

Third Floor
三层平面

PROJECT SIX : SKY PARK CHAPTER 8

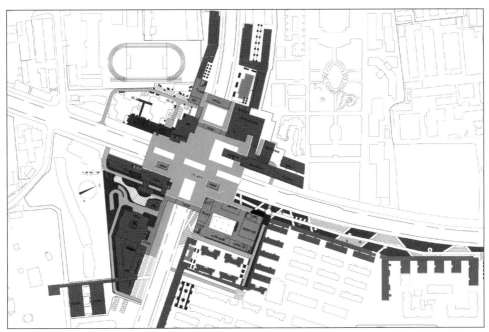

Second Floor
二层平面

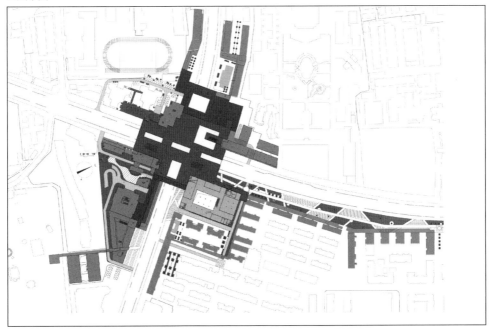

Fourth Floor
四层平面

作品六：空中公园

PEDESTRIAN PARADISE ON ARTERIAL ROAD

SIPING ROAD - QUYANG ROAD
四平路—曲阳路

ART RIVER
艺术河

Peng Hedong / Anna Diallo   彭何东/迪安娜

The beauty of this place is coming from the river, which is flowing between the buildings, bringing some culture into an empty place. We noticed that the lack of space in the riverside low-rise houses residents makes them gather outside in the streets. That's why instead of destroying those important buildings, some open plazas are now occupying some old busy places. The north side with its huge exhibition center emphasizes the main function of the whole program, providing art to the residents. Playing with the two sides of the river the main principle of the design is to use both old and new buildings to create a global artistic atmosphere. Finally the situation, very close to a metro station, is an occasion to enhance this important crossroad of Shanghai. The pedestrian path proposed going up and down, under earth under the river, punctuated by activities and views becomes a real attraction.

四平路与曲阳路的交叉口是虹口港水系与干道交汇的地方，河流以及沿岸低层建筑给这个地方带来了别具一格的城市意象，这是我们设计需要珍惜的。设计力图利用新旧建筑共同来创造一个国际化的艺术氛围，而此地离地铁站很近的优势，又将最终把这个重要的十字路口推向整个上海。人行流线被设计成上下多个层次，有地下、河下的部分，注入活动后会变得极富吸引力。我们希望有更多可能的艺术之河能在上海创造更多的城市聚集点。滨河步行路径和穿过机动车道路的下穿通道形成了完整的体系，同时把机动车的影响排除在外。

BIRDVIEW ON SIPING ROAD-QUYANG ROAD NODE
四平路—曲阳路节点鸟瞰

Research Scope of Siping Road-Quyang Road Node
四平路—曲阳路节点研究区域

PROJECT SEVEN: ART RIVER   CHAPTER 8

作品七：艺术河

PEDESTRIAN PARADISE ON ARTERIAL ROAD

Ground floor plan and circulation 一层平面及环线

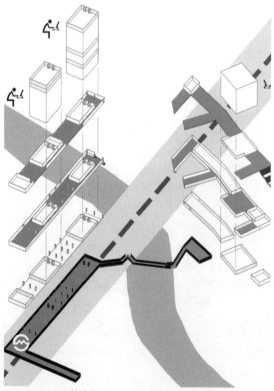

Open Area 开放空间

PEDESTRIAN PARADISE ON ARTERIAL ROAD

⑧ SIPING ROAD - LINPING NORTH ROAD
四平路—临平北路

DEFRAGMENTATED SPACES
去碎片化的城市空间

Li Hengye / Clarisse Carlier 李恒晔/丽斯

The core problems in the joining point of Linping Road and Siping Road lies in that the situation of traffic jam in this node is very serious. The high-density development on both side of arterial roads doesn't leave enough space for pedestrian space, while the border to the sidewalks is not design with open façade which fails to provide urban functions to the sidewalks. Besides, the building properties adjacent to the node are almost self-enclosed and isolated to each other, due to which the pedestrian net does not exist. Our urban Design makes research on possibilities for these properties to open up their site, active rest spaces within their sites and links them together into a pedestrian-friendly and vivid urban space net. At last, the improved pedestrian space alongside Siping Road will act as a center to the cross-road public space net.

　　临平路与四平路的交叉点的核心问题在于这里的交通拥堵非常严重，两侧建筑高密度开发并没有对沿街步行空间进行考虑，界面开放不足，功能欠缺。此外，周边的地块开发全部是封闭式的，每个地块相互隔离，没有形成公共空间的网络。我们的城市设计研究了周边区域进行开放的可能性，设计将这些地块内的剩余空间激活并彼此相连，形成一个完整的适宜步行的开放空间网络，优化后的四平路两侧人行道成为跨路网络连接的中枢地带。

BIRDVIEW ON SIPING ROAD-LINPING NORTH ROAD NODE
四平路—临平北路节点鸟瞰

Research Scope of Siping Road-Linping North Road Node
四平路—临平北路节点研究区域

PEDESTRIAN PARADISE ON ARTERIAL ROAD

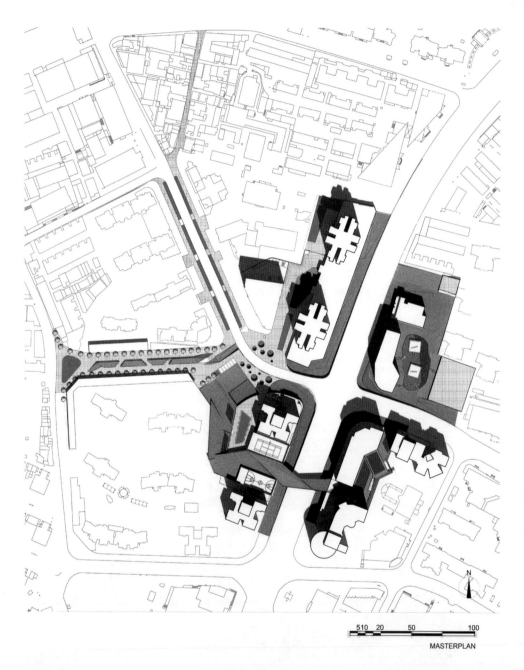

Master Plan
总平面

PROJECT EIGHT: FRAGMENTAL SPACES CHAPTER 8

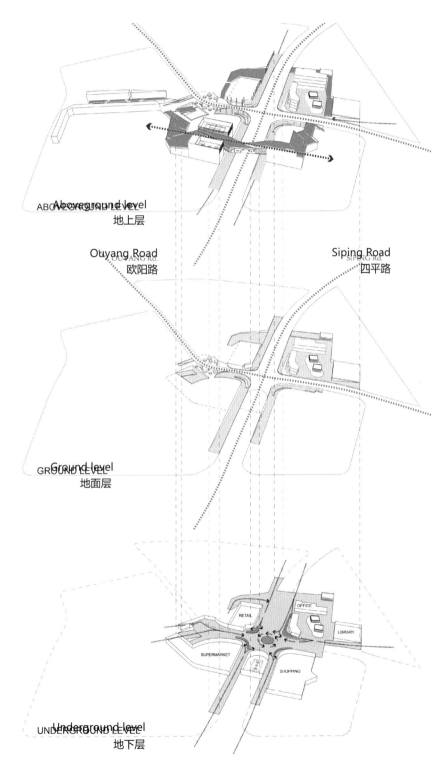

Aboveground level
地上层

Ouyang Road
欧阳路

Siping Road
四平路

Ground level
地面层

Underground level
地下层

PEDESTRIAN PARADISE ON ARTERIAL ROAD

⑨ SIPING ROAD - HAILUN ROAD
四平路—海伦路

GREEN LINK
绿色纽带

Cai Xingjie / Christoph Holzinger / Yang Dan   蔡兴杰/霍雅/杨丹

Residential buildings, Lilong residences, a five-star hotel and a construction site are the main surroundings substances of Siping Road – Hailun Road Node. Metro Line 4 and Line 10 meet here. Complicated urban fabric and busy city traffic are the two considerable characters of Hailun Road, while it is not a comfortable place to stay because of the lack of public space and walkable areas. Hence, creating more diverse public space and a friendly pedestrian system is our main goal to reach.

Zooming out our design scope, by footbridges combined with functions and landscapes to connect North Sichuan Road Park, Hailun Park and our city complex design, we attempt to generate a whole new pedestrian system and a noticeable increase of public space. The linear park which is almost 2 km long, will be the new spatial core (in the whole region) ,which stimulates the further development of the linear-shaped area alongside Siping Road.

居民区、里弄、一个五星级酒店和一个工地是四平路—海伦路的主要周边构成，轨道交通4号线和10号线也汇聚于此。复杂的城市肌理和繁忙的城市交通是海伦路最大的两个特点；然而，公共空间和步行区域的缺乏，使得这里并不适合停留，因此创造出更丰富的公共空间和更人性化的步行系统是我们最主要的设计目标。

我们把设计的视野放大，通过功能与景观结合的步行桥将四川北路公园、海伦公园与我们新设计的城市综合体串联，形成一个整体的步行系统同时也创造出更多的公共空间。这个长达2公里的城市景观带将会成为整个区域公共空间网络的主轴，并激发四平路沿线城区的进一步发展。

BIRDVIEW ON SIPING ROAD-HAILUN ROAD NODE
四平路—海伦路节点鸟瞰

Research Scope of Siping Road-Hailun Road Node
四平路—海伦路节点研究区域

PEDESTRIAN PARADISE ON ARTERIAL ROAD

Roads 道路

Rivers 河流

Metro Exits 地铁出口

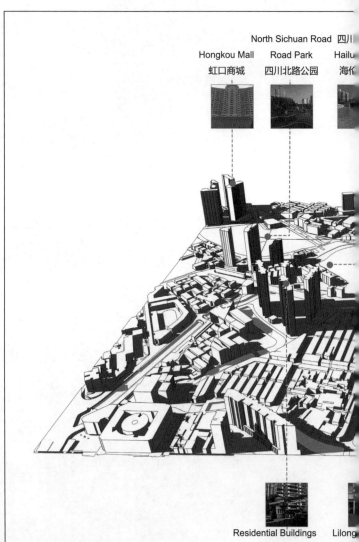

PROJECT NINE: GREEN LINK   CHAPTER 8

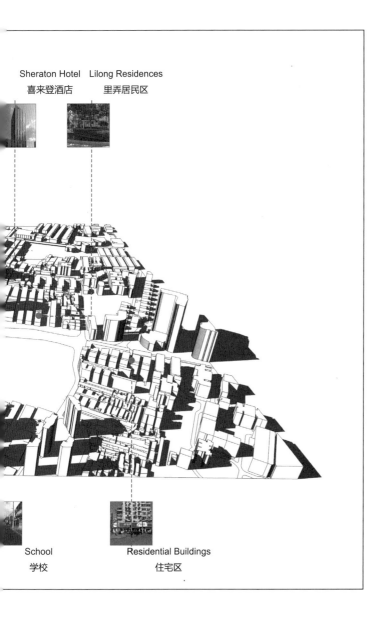

Sheraton Hotel　Lilong Residences
喜来登酒店　里弄居民区

School　　　　Residential Buildings
学校　　　　　住宅区

Residences 住宅

Builings over 10 floors
10层以上建筑

Builings over 100m
100米以上建筑

PEDESTRIAN PARADISE ON ARTERIAL ROAD

North Sichuan Road Park
四川北路公园

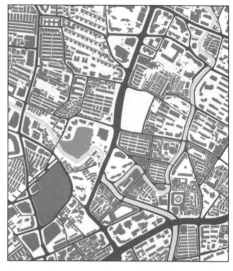

Hailun Park Reactivation
海伦公园

New Park
新的公园

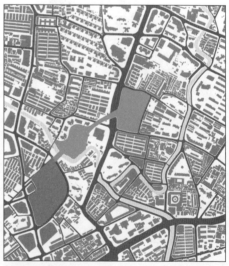

City Complex Design
城市综合体

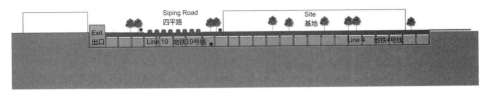

Section
剖面图

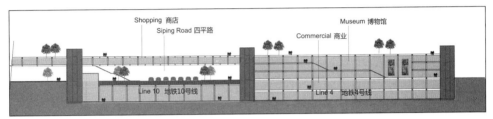

Section
剖面图

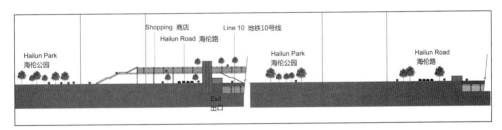

Section
剖面图

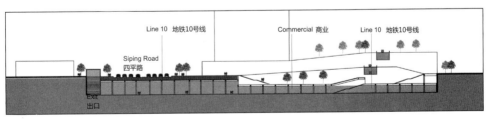

Section
剖面图

Sections 剖面图

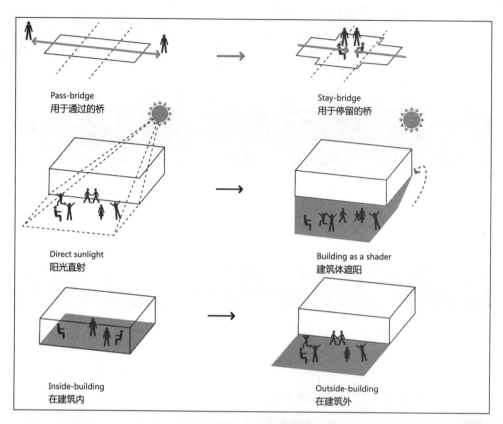

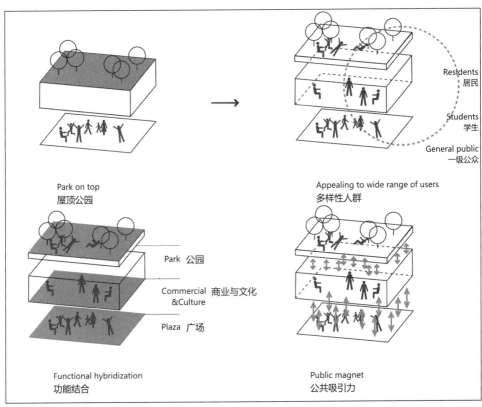

Bridge Diagrams 桥的图解

PEDESTRIAN PARADISE ON ARTERIAL ROAD

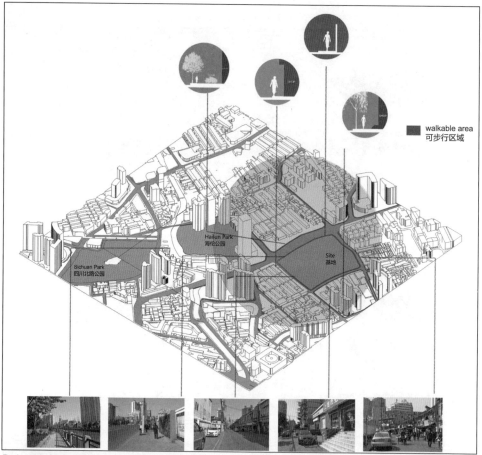

walkable area
可步行区域

Pedestrian System Before
已有的步行系统

PROJECT NINE: GREEN LINK   CHAPTER 8

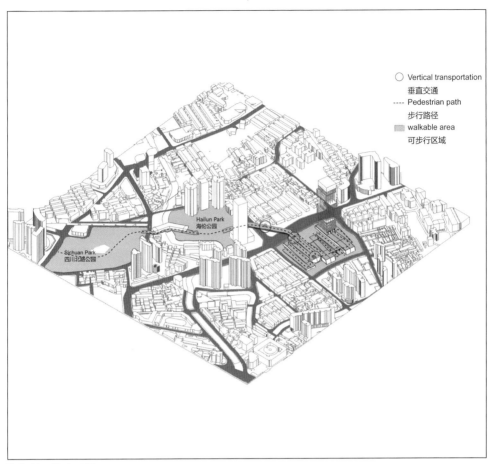

Pedestrian System After
新的步行系统

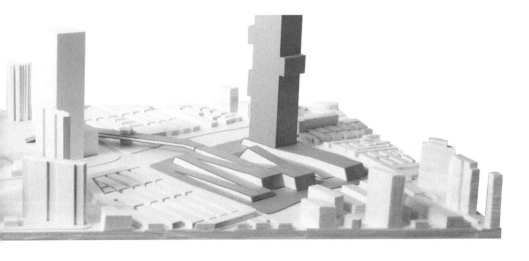

PEDESTRIAN PARADISE ON ARTERIAL ROAD

⑩ WUSONG ROAD - HAINING ROAD
吴淞路—海宁路

TOWARD THE INTERSECTION
"天桥地带"

Zhu Feng / Arnaud Despretz   朱枫/阿诺

Wusong Road Haining Road is an important traffic node of North Bund, its importance can not be ignored, but how to make space become rich and vivid? Our process is designed around the basis of the status quo of Haining Road and Siping Road. We try to make people stay and find something interesting, to discover and appreciate the process in pass by the annular Bridge. The bridge is designed with certain urban function, in which the usage of bridge will turn from pure transportation activity to an urban experience. This syetem is further extended to both sides into Haining Road's sidewalk space to activate space there, which serves as a starting point of street net regeneration in this area.

吴淞路海宁路作为北外滩一个重要的交通节点，其重要性不可忽略；但是，如何通过设计使这个仅限于人们通行的节点空间变得丰富生动则是我们探讨的话题。我们围绕四平路和海宁路的现状，进行调研，分析和重建，使人们在路过这个环形天桥的时候有一个驻足停留、发现和观赏的过程。环形天桥设置一定的功能让跨路的活动脱离单纯的功能性，而变成一种城市体验。这个系统进一步向两侧延伸，令海宁路店沿线的空间生动起来，并向纵深方向生长，街道的网络得以慢慢形成。

BIRDVIEW ON WUSONG ROAD-HAINING ROAD NODE
吴淞路—海宁路节点鸟瞰

Research Scope of Wusong Road-Haining Road Node
吴淞路—海宁路节点研究区域

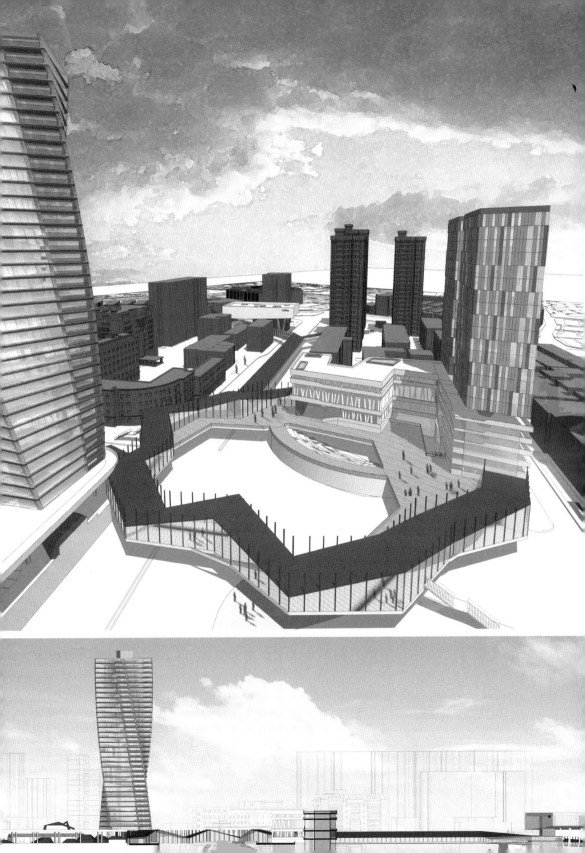

PEDESTRIAN PARADISE ON ARTERIAL ROAD

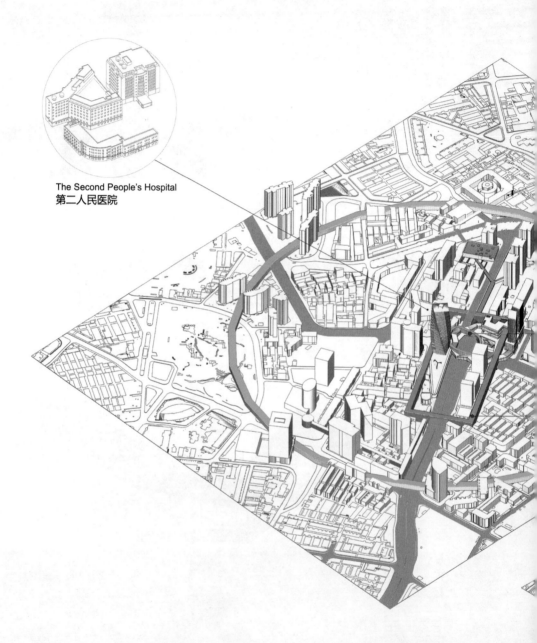

The Second People's Hospital
第二人民医院

PROJECT TEN: TOWARD THE INTERSECTION   CHAPTER 8

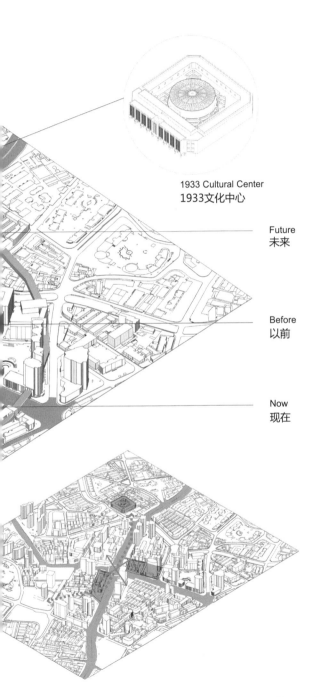

1933 Cultural Center
1933文化中心

Future
未来

Before
以前

Now
现在

The intersection of existing pedestrian overpass has single function, when people pass through, few are willing to stay. Our design is aimed at changing the status quo. In this design, we expand the radius of the annular bridge link, which is combined with commercial, square and activities.

路口现有的步行天桥只有单一的使用功能，人们来往穿行，很少有人愿意驻足停留，我们的设计旨在改变这一现状。在总体设计上，我们扩大了环形天桥的链接半径，将商业、广场、活动融入其中。

Concept 概念生成

作品十："天桥地带"   227

PEDESTRIAN PARADISE ON ARTERIAL ROAD

The name of the design is "toward the intersection". In the beginning, based on the survey we have done, that most blocks are separated from each other directly, the connection of the circle bridge is weak.Thus, we enlarge the Circle Bridge and make the pedstrian system stretch into the buildings . Also we created more degrees of space to contain various kinds of activities from the private to the public.

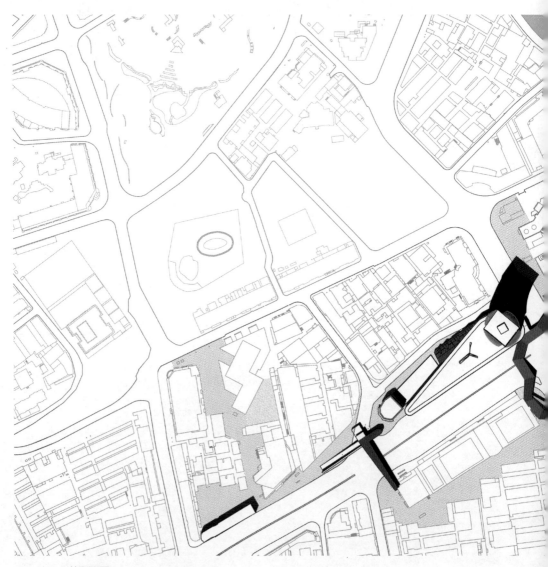

Master Plan 总平面图

之所以将我们的设计取名为"环形交往组织",是因为在调研初期,我们发现环形天桥非常缺乏活力和生机,它对人们行为的干涉几乎为零,所以我们的设计扩大了环形天桥的辐射范围,同时设置许多开放空间和广场。人们在从一个点到达另外一点穿行的时候不再显得这么单调乏味,开放空间反应在竖向上的关系也更为多样。

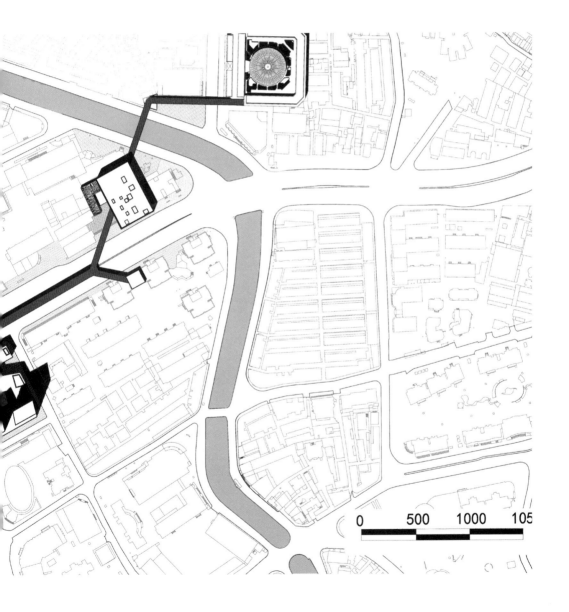

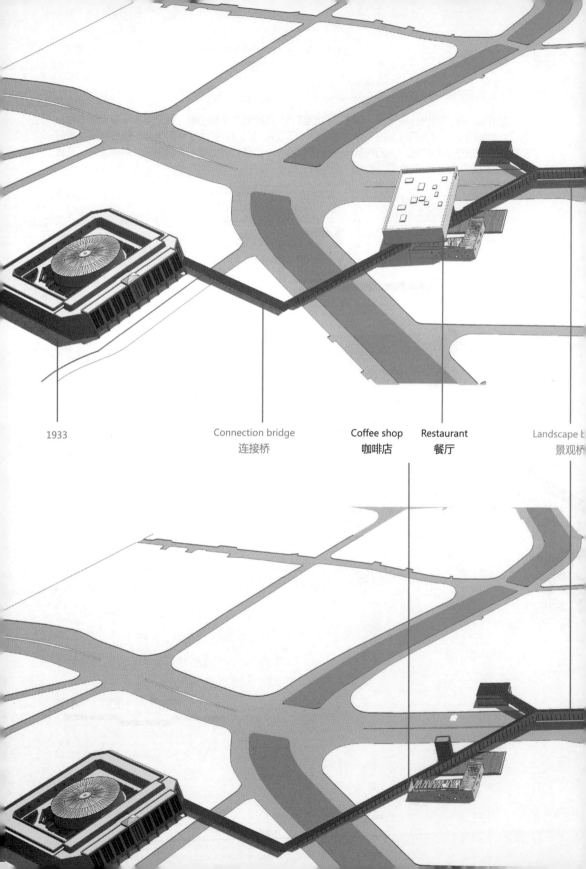

1933　　　Connection bridge　　Coffee shop　Restaurant　　Landscape b
　　　　　连接桥　　　　　咖啡店　　餐厅　　　　景观桥

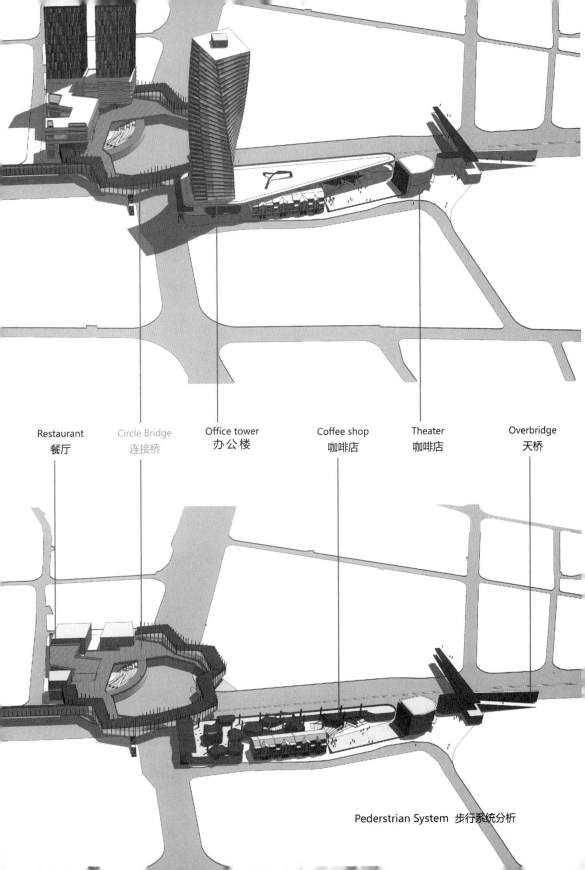

Restaurant 餐厅　Circle Bridge 连接桥　Office tower 办公楼　Coffee shop 咖啡店　Theater 咖啡店　Overbridge 天桥

Pederstrian System 步行系统分析

⑪ WUSONG ROAD - THE BUND
吴淞路—外滩

PEDESTRIAN PARADISE FACING SUZHOU CREEK
面向河流的步行天堂

Li Jing / Lisa Mueller / Chen Yantong / Liu Hongxi　李京/穆丽莎/陈彦彤/刘泓汐

The site is located right at the important crossing of Huangpu River and the Suzhou Creek. It is also where Rockbund, Bund, Waibaidu Bridge, Broadway Mansions and many other important historic buildings are located. This design attempts to create an pedestrian area, that forms a counterpart to the Bund, by creating an attractive public space for inhabitants, workers and tourists alike and offering them a multifunctional, dynamic area to enjoy their time. Different platforms are offering a wide variety of seating options, on which people can sit like on a stage, enjoying the view and the waterfront. Two towers, that include a library and an art centre for the public, but also housing and offices, are completing the new skyline of Suzhou creek. Traffic here is reduced to make way for public space.

　　场地是苏州河和黄浦江之间的一个重要节点。在这里有外滩源，外滩和外白渡桥，百老汇大厦和许多重要的历史建筑。经过深入的分析，一些主要的问题可以被发现：缺少可用的公共空间和缺少亲水空间。这个设计意在对这些问题做出应对，通过为居民创造一个有吸引力的、功能复合、充满活力的空间以令他们尽享休闲时光，更重要的是吸引他们的停留。不同的平台提供了丰富的座椅选择。两座塔楼的功能包括图书馆和艺术中心，也有一些住宅和办公建筑，它让苏州河的天际线完整，并形成了一个从百老汇大厦到另一边场地的联系。这段道路的机动车流量不大，道路周边步行环境的提升拥有天然的条件。

BIRDVIEW ON WUSONG ROAD-THE BUND NODE
吴淞路—外滩节点鸟瞰

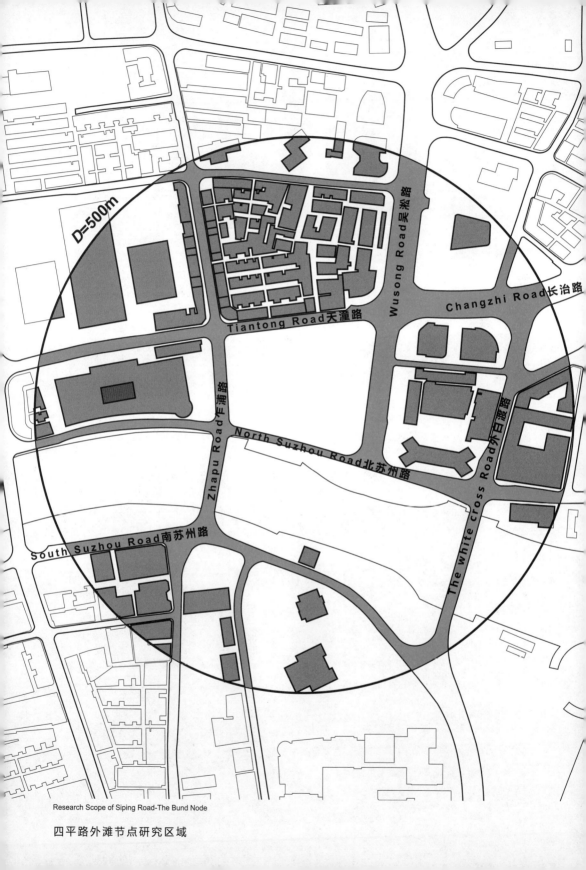

Research Scope of Siping Road-The Bund Node

四平路外滩节点研究区域

PEDESTRIAN PARADISE ON ARTERIAL ROAD

THE SITE 选址

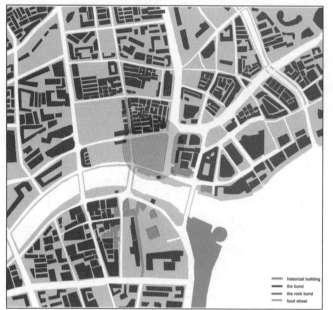

Site
基地

Functions
功能

Height Analysis
高度分析

Walkability
步行范围

CURRENT TRAFFIC SYSTEM 现状交通系统

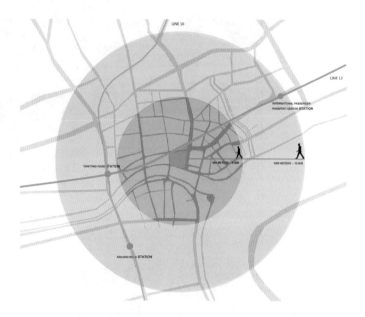

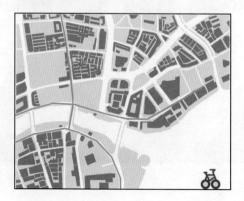

SECTIONS 剖面

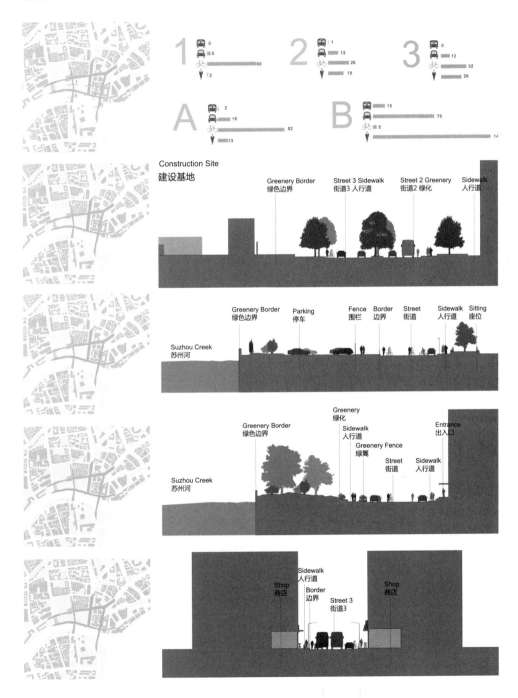

PEDESTRIAN PARADISE ON ARTERIAL ROAD

PROJECT ELEVEN: PEDESTRIAN PARADISE FACING SUZHOU CREEK   CHAPTER 8

References 参考文献

1. Addison, C., Zhang, S. and Coomes, B. Smart growth and housing affordability: A review of regulatory mechanisms and planning practices[J]. Journal of Planning Literature 2013 (28-3):215–257.
2. Talen, E. and Koschinsky, J. The walkable neighborhood:A literature review[J]. International Journal of Sustainable Land Use and Urban Planning 2013 (1-1): 42–63.
3. Bauman A E, Reis R S, Sallis J F, et al. Correlates of physical activity: why are some people physically active and others not?[J]. The lancet, 2012, 380(9838): 258-271.
4. Duncan D T, Aldstadt J, Whalen J, et al. Validation of Walk Scores and Transit Scores for estimating neighborhood walkability and transit availability: a small-area analysis[J]. GeoJournal, 2013, 78(2): 407-416.
5. Forsyth A. What is a walkable place? The walkability debate in urban design[J]. Urban Design International, 2015, 20(4): 274-292.
6. Forsyth, A., & Southworth, M. Cities afoot- Pedestrians, walkability, and urban design[J]. Journal of Urban Design, 2008 (13-1) : 1–3.
7. Frank L D, Engelke P. Multiple impacts of the built environment on public health: walkable places and the exposure to air pollution[J]. International Regional Science Review, 2005, 28(2): 193-216.
8. Gebel K, Bauman A, Owen N. Correlates of non-concordance between perceived and objective measures of walkability[J]. Annals of behavioral medicine, 2009, 37(2): 228-238.

9. Gehl J. Life between buildings: using public space[M]. Island Press, 2011.
10. Jacobs, J. The Death and Life of Great American Cities [M]. New York: Random House. 1961.
11. Moudon A V, Lee C, Cheadle A D, et al. Operational definitions of walkable neighborhood: theoretical and empirical insights[J]. Journal of Physical Activity & Health, 2006, 3: S99.
12. Owen N, Humpel N, Leslie E, et al. Understanding environmental influences on walking: review and research agenda[J]. American journal of preventive medicine, 2004, 27(1): 67-76.
13. Sallis J F, Frank L D, Saelens B E, et al. Active transportation and physical activity: opportunities for collaboration on transportation and public health research[J]. Transportation Research Part A: Policy and Practice, 2004, 38(4): 249-268.
14. Southworth M. Designing the walkable city[J]. Journal of urban planning and development, 2005, 131(4):246-257.
15. Speck J. Walkable city: How downtown can save America, one step at a time [M]. Macmillan, 2013.
16. Sugiyama T., Neuhaus, M., Cole, R., Giles-Corti, B. and Owen, N. Destination and route attributes associated with adults'walking: A review[J]. Medicine and Science in Sports & Exercise 2012 (44-7): 1275–1286.
17. Whyte W H. The social life of small urban spaces[M]. 1980.

后记

本书是课题组近三年来关于城市步行系统研究的一次展示。在国家自然科学基金的资助下，课题组老师、研究生都付出了大量精力，同时也收获了丰硕的成果，在此向参加基金项目研究的各位老师和同学表示感谢。感谢同济大学建筑与城市规划学院的刘刚副教授为本书关于四平路的发展和历史沿革提供了详细的资料并撰写了部分重要内容。感谢维也纳工大的克劳斯·森姆斯罗特（Klaus Semsroth）教授和莫拉登·雅德里奇（Mladen Jadric）副教授，感谢他们对于本研究的协同探索和许多精彩的想法，他们不但每个学期都参与了设计课程的评图与指导，克劳斯·森姆斯罗特教授还为课程进行了两场非常重要的讲座。感谢比利时布鲁塞尔自由大学的马可·伦赞尔托（Marco Renzato）副教授，受邀于2014年春季学期，参与了"步行者天堂"的课程设计指导。感谢韩国釜山大学的李仁熙教授、同济大学的王一副教授、王志军副教授、俞泳副教授、哈利·邓·哈托格（Harry den Hartog）助理教授、清华大学的续倩博士等对本课程的点评和讨论。

在本书的编辑和排版过程中，维也纳工大的双学位硕士研究生丽莎·穆勒（Lisa Müller）同学为城市设计实践案例的图纸整理和编排做了大量工作，同济大学硕士研究生段翔宇同学、朱薛景同学在文字编辑和排版上做出了重要贡献，研究生张家洋、杜叶铖、李曼竹、陈梦梦、王登恒等同学做了城市干道的相关调研工作，并绘制了部分插图，在此一并表示感谢！

Epilogue

The research and publication is supported by NSFC. The book is a joint effort of the author and many other professors and students. Special thank is given to Prof. Liu Gang of Tongji University, who takes his time to contribute in important part of this book regarding the history evolution of Siping Road. Gratitude is also sent to Prof. Klaus Semsroth and Prof. Mladen Jadric from Vienna University of Technology who contributed with many ideas for the research and gave student wonderful lectures regarding urban space, to Prof. Marco Renzato took part in the teaching of studio in 2014, and Prof. LEE Inhee from Pusan National University ,Prof. Wang Yi, Prof. Wang Zhijun, Prof. Yu Yong and Prof. Harry den Hartog from Tongji University as well as Dr. Xu Qian from Tsinghua University.

The publication wouldn't be completed without efforts of our master students Lisa Müller (Austria), Duan Xiangyu, Li Manzhu and Zhu Xuejing, who devoted themselves for the editing and graphics of the book. Zhang Jiayang, Du Yecheng, Li Manzhu, Chen Mengmeng and Wang Dengheng contributed a lot with their case investigation and drew some figures in this book.

图书在版编目（CIP）数据

步行与干道的合集：中英对照 / 孙彤宇, 许凯著.
--上海：同济大学出版社, 2017.4
ISBN 978-7-5608-6677-2

Ⅰ.①步... Ⅱ.①孙... ②许... Ⅲ.①城市道路 - 城市规划 - 研究 - 中国 - 汉、英 Ⅳ.①TU984.191
中国版本图书馆CIP数据核字(2016)第308505号

城市设计研究
步行与干道的合集
【著】孙彤宇　（同济大学）
　　　许　凯　（同济大学）

责任编辑　武　蔚
责任校对　徐春莲
版面设计　陈梦梦　朱薛景
出版发行　同济大学出版社　www.tongjipress.com.cn
　　　　　（地址：上海市四平路1239号　邮编：200092　电话：021-65985622）
经　　销　全国各地新华书店
印　　刷　常熟市华顺印刷有限公司
开　　本　787mm×960mm　1/16
印　　张　16
字　　数　320 000
版　　次　2017年4月第1版　2017年4月第1次印刷
书　　号　ISBN 978-7-5608-6677-2
定　　价　68.00元

本书如有印装质量问题，请向本社发行部调整　版权所有　侵权必究